CAMBRIDGE TRACTS IN MATHEMATICS

General Editors

B. BOLLOBAS, P. SARNAK, C.T.C. WALL

100 **Analysis and Geometry on Groups**

Analysis and Geometry
on Groups

N. Th. Varopoulos
L. Saloff–Coste
T. Coulhon

Université de Paris, VI

 CAMBRIDGE
UNIVERSITY PRESS

CAMBRIDGE UNIVERSITY PRESS
Cambridge, New York, Melbourne, Madrid, Cape Town, Singapore, São Paulo, Delhi

Cambridge University Press
The Edinburgh Building, Cambridge CB2 8RU, UK

Published in the United States of America by Cambridge University Press, New York

www.cambridge.org
Information on this title: www.cambridge.org/9780521353823

First published 1992
This digitally printed version 2008

A catalogue record for this publication is available from the British Library

ISBN 978-0-521-35382-3 hardback
ISBN 978-0-521-08801-5 paperback

CONTENTS

PREFACE

Many things could be said about the way this book was written but we shall be brief.

It all started with several lecture courses given by N. Varopoulos at Université Paris VI during the period 1982-87. At the time, Coulhon and Saloff-Coste were post-doctoral students and took notes. An early part of these notes appeared for limited circulation in 1986. It was then decided that, when completed, these notes would be published as a set of graduate "Lecture Notes". The project dragged on for several years; by 1990, through the efforts of Saloff-Coste, enough work had been put into the notes to make them presentable as a real book.

This book is primarily an advanced research monograph. It should be accessible to those graduate students that are prepared to make the personal investment and effort to familiarize themselves with the background material.

N. Varopoulos did very little of the actual writing and did not put any work into the preparation of manuscripts; he is however responsible for most of the new mathematics that is presented here. This mathematical work was done during the 1980s and was built on the following basic material.

Existing semigroup theory, especially Beurling–Deny theory; this is work that was done in the 1950s and 1960s. The work of J. Moser and J. Nash on parabolic equations was also a great inspiration in this context.

The theory of second order subelliptic differential operators and especially the "sum of squares operators". This is work done in the 1960s by L. Hörmander. The Harnack estimates, which are essential for us, were completed by J.-M. Bony a little later. This work has since been further developed by several authors.

Finally, the basic real analysis that we all know and which has its origins in the work of Hardy, Littlewood, Marcinkiewicz, etc. in the 1930s.

In fact, *grosso motto*, the above points and all that goes around them, are the background material to which I referred earlier. Some of it is explained but a beginner will no doubt find our explanations a bit concise. To make all this background really accessible to such a reader would have more than doubled the size of the book, and anyway, none of us was prepared to do it!

Together with the two laboratories in our own University, namely the U.A. 213 and the U.A. 754, we wish to thank the following institutions, that, during the preparation of the book, have offered hospitality and support to one or more of us: University of California, Los Angeles (USA); MIT, Cambridge (USA); McGill University (Canada); Mittag–Leffler Institute (Sweden); Institute of Mathematics of Wroclaw University (Poland).

We also have to thank the following colleagues who have read parts of the book in manuscript form, and sometimes as well were kind enough to make corrections, and in addition discussed with us aspects of the book (mathematical or otherwise) while we were in the process of writing: A. Ancona, E.B. Davies, D.J.H. Garling, D. Lamberton, M. Ledoux, N. Lohoué, P. Maheux, Ph. Mongin, and D. Stroock.

Thanks are due to R. Bruno who had a big influence on the decisive early phase of the research that led to this book.

Finally this book owes a great deal to Cambridge University Press and to its Mathematics Editor, David Tranah. He combines two qualities that very seldom go together: he is very professional in his job and at the same time very relaxed in human relations. It was a pleasure doing business with him.

N. Th. Varopoulos
Paris, April 1, 1992

FOREWORD

This short foreword has two purposes. First of all it will clarify the title of the book which might otherwise be deemed a trifle pretentious. It will also give in a few lines some of the motivating problems and a brief history of the subject over the last ten years that led us to write this book.

Considering groups as geometric objects is an old idea. Indeed, one could say that Lie groups are, after all, more than anything else geometric objects. More recently topologists have made good use of this idea to give geometric proofs of purely algebraic theorems. From our point of view however the starting point for the geometric considerations on groups that we shall be interested in can be traced to a paper by Milnor, from 1968. Milnor showed that the fundamental group of a negatively curved compact manifold is of *exponential growth*.

Let us stop for a minute to discuss this theorem. The fundamental group of any compact manifold is a finitely generated discrete group, G. Our first task is to define a natural distance on such a group. Let G be generated by $g_1, ..., g_k \in G$ and let us decree that the ball of radius n centred at $e \in G$ is exactly the set of group elements of the form $g = g_{i_1}^{\varepsilon_1} \cdots g_{i_p}^{\varepsilon_p}$ ($0 \leq p \leq n$, $i_\alpha = 1, 2, ..., k$, $\varepsilon_\alpha = \pm 1$). Using these balls we can define the distance from e, $d(x, y)$ ($x, y \in G$) by left translation on G. The number of elements in the above n-ball is denoted by $\gamma(n)$ and referred to as the *growth function* of G (Milnor's theorem asserts that the fundamental groups that he considers satisfy $\gamma(n) \geq e^{cn}$.)

After Milnor's paper the volume growth of a group became a subject of research and in 1981 M. Gromov, in a famous paper, gave an algebraic characterization of the groups that satisfy $\gamma(n) = O(n^A)$ (for some fixed $A \geq 0$).

The next natural geometric question to consider is the isoperimetric inequality on finite subsets $\Omega \subseteq G$ of a (finitely generated discrete) group. Let us define

$$\partial \Omega = \{\omega \in \Omega \mid d(\omega, G \setminus \Omega) \leq 1\}$$

and, for some $A \geq 1$, examine the possible validity in G of an inequality of the form:

$$|\Omega|^{\frac{A-1}{A}} \leq C|\partial\Omega|; \quad \Omega \subseteq G \qquad \qquad \text{Iso}(A):$$

where $|\cdot|$ denotes the cardinality of the set and $C > 0$ is independent of Ω. It is very easy to see that if Iso(A) holds in G then $\gamma(n) \geq C_1 n^A$. (Indeed, Iso(A) implies on $\gamma(t)$ the differential inequality

$$\frac{\mathrm{d}}{\mathrm{d}t}\gamma(t) \geq C^{-1}\left(\gamma(t)\right)^{\frac{A-1}{A}}$$

which gives the result.) One of the theorems that will be proved in this book says that the converse statement also holds.

Let us be more explicit and, for every $f \in c_0(G)$, define the ℓ_p-norm of the gradient

$$||\nabla f||_p^p = \sum_{d(x,y) \leq 1} |f(x) - f(y)|^p .$$

It is very easy to see that Iso(A) is equivalent to

$$||f||_{\frac{A}{A-1}} \leq C ||\nabla f||_1; \quad f \in c_0(G). \qquad\qquad \text{Sob}(A):$$

One of the main aims of this book is therefore to prove the following geometric result:

$$\text{Sob}(A) \quad \Longleftrightarrow \quad \gamma(n) \geq Cn^A \quad n = 1, 2, \ldots$$

We shall also be concerned with analysis and potential theory on groups. Again one could say that this is a very old subject, indeed $\mathbb{R}^n$ is after all a group. More recently (and more significantly from our point of view) an extensive study of an important class of nilpotent groups has been undertaken by many analysts and basic applications to P.D.E.s and complex analysis were discovered. At more or less the same time probabilists and potential theorists have been examining random walks and potential kernels on groups.

If $\mu \in \mathbb{P}(G)$, for any probability measure on, say the discrete, group G one can consider the random walk on $\{Z_n \in G\}_{n \geq 1}$ defined by

$$\mathbb{P}[Z_n = g \,//\, Z_{n-1} = h] = \mu(g^{-1}h).$$

One can then ask all the natural questions. For instance, is this walk transient or recurrent? This is usually referred to as Kesten's problem.

A closely related problem can also be considered. Let M be a Riemannian manifold and assume that M is a normal covering of some compact manifold K. We denote by G the deck transformation group so that $M/G \cong K$. This is the same set-up that we had for Milnor's theorem. What started the research in this book was a theorem (proved in the early 1980s) which stated that the canonical Brownian motion on M is transient (respectively, recurrent) if and only if the random walks of Kesten's problem on G are transient (respectively, recurrent), (for appropriate symmetric measures $\mu \in \mathbb{P}(G)$).

Indeed this theorem showed that there was an intimate connection between discrete combinatorial considerations such as a random walk on a discrete group or a graph, and a continuous set-up where analytic tools are more effective. The way to prove the above theorem is to identify G with a discrete "skeleton" of M and use the "natural discretization" of the problem. The correct framework for that "natural discretization" is the Beurling-Deny

One feels therefore that it ought to be possible to express $\phi(n)$ $(n \to \infty)$ by some group invariant. It is reasonable to conjecture that

$$\phi(n) \sim [\gamma(\sqrt{n})]^{-1},$$

which is unfortunately only correct if $\gamma(n)$ grows polynomially. Alternatives to this conjecture will be examined later on.

One final aspect of the theory that should be mentioned evolves around the Gaussian, off-diagonal, estimates. These are estimates of the form:

$$\mu^{*n}(g) \leq C^{-1} \exp\left(-C\frac{|g|^2}{n}\right).$$

These estimates are significant when $|g| = d(e, g)$ is large, and they are important if we want to have a complete picture of what is happening. The ideas of E. B. Davies are vital for these estimates. (These ideas have been presented in a recent monograph in the same series.)

The geometric and the analytical aspects of the theory that we have described above are held together not only by conceptual considerations but also by the methods of the proofs. Indeed these proofs are based on both aspects simultaneously and it would be impossible to give them without keeping the two sides in mind all along.

There are many topics closely related to our subject that we have not touched upon at all in this book. For example, Riesz transforms, non-unimodular heat kernels, homogeneous spaces, connections with symmetric spaces, and so on.

To finish on an optimistic note, let us express the hope that these topics, which are now in full development, might one day find their place in another book.

theory of Dirichlet spaces. Shortly afterwards the Beurling-Deny theory became fundamental for the whole subject. Indeed, from another point of view, the transience criterion of Beurling and Deny (applied directly to the semigroups of the random walks on G) implies the following result:

Let $\mu \in \mathbb{P}(G)$ be symmetric and assume that it has finite generating support. Then the corresponding random walk on G is transient if and only if the following *a priori* inequality holds:

$$|f(e)| \leq C \cdot ||\nabla f||_2; \quad f \in C_0(G). \qquad \text{(Tran.)}:$$

Observe that the criterion (Tran.) is independent of the particular choice of μ, and only depends on the "geometry of G".

If we compare (Tran.) with (Sob) we see that our geometric theorem implies that all groups for which $\gamma(n) \geq Cn^{2+\varepsilon}$ are transient. This, combined with Gromov's theorem, finally yields that the only groups which are not transient are the finite extensions of $\{0\}$, $\mathbb{Z}$, $\mathbb{Z}^2$.

A critical reader could argue that the above problems are combinatorial and that we have not really been doing analysis but discrete mathematics in disguise. To convince such a reader, let us push the process a stage further and examine the asymptotic behaviour of the convolution powers $\mu^{*n}(e)$ for $\mu \in \mathbb{P}(G)$ as above. The transience or recurrence of the random walk can, after all, be determined from the convergence or divergence of the series $\sum \mu^{*n}(e)$.

For the measures that we have been considering, $\mu^{*n}(e)$ is the $\ell^1 \to \ell^\infty$ convolution operator norm of μ^{*n} and the behaviour of this is, for large n, equivalent to the behaviour of $||T_t||_{1 \to \infty}$ where

$$T_t = \exp(-t(\delta - \mu))$$

is the continuous time semigroup attached to the walk. What is central here is the following (functional) analytic result:

$$||T_t||_{1 \to \infty} = O(t^{-A/2}) \iff ||f||_{\frac{2A}{A-2}} \leq C||\nabla f||_2; f \in c_0(G)$$

valid for any $A > 2$ and for very general semigroups. This theorem shows the intimate connection that exists between Sobolev inequalities and geometry on the one hand, and the analysis of the asymptotic behaviour of natural semigroups on the other.

Many such natural semigroups can be constructed on Lie groups starting from their infinitesimal generators Δ that are second order left-invariant subelliptic differential operators on the group. To close this circle of ideas, observe finally that the heat diffusion semigroup $e^{-t\Delta}$ on $\mathbb{R}^n$ or more generally on a nilpotent Lie group is the main building block for the real analysis there.

Another consequence of the above theorem is that the behaviour of $\phi(n) = \mu^{*n}(e)$ (as $n \to \infty$) measured on the polynomial scale, is independent of μ.

CHAPTER I

INTRODUCTION

I.1 Sobolev inequalities in $\mathbb{R}^n$

One of the aims of this book is to study Sobolev inequalities on Lie groups. It is thus natural to present the situation in the simple case of $\mathbb{R}^n$.

In 1936, S. Sobolev published a paper in which he proved a host of *a priori* inequalities. He showed in particular that

$$\|f\|_{\frac{np}{n-p}} \leq C\|\nabla f\|_p, \quad \forall f \in C_0^\infty(\mathbb{R}^n), \tag{1}$$

for all $1 < p < n$. It is easy to see that the inequality

$$\|f\|_q \leq C\|\nabla f\|_p, \quad \forall f \in C_0^\infty(\mathbb{R}^n), \tag{2}$$

cannot hold unless $p < n$ and $q = np/(n-p)$. Indeed, if one replaces f by $f_\lambda : x \mapsto f(\lambda x)$, $\lambda > 0$ in (2), one gets

$$\lambda^{-n/q}\|f\|_q \leq C\lambda^{1-n/p}\|\nabla f\|_p, \quad \forall f \in C_0^\infty(\mathbb{R}^n).$$

This forces the above conditions.

Sobolev's paper does not contain a proof of the case $p = 1$. However, on any Riemannian manifold, the inequality

$$\|f\|_q \leq C\|\nabla f\|_1, \quad \forall f \in C_0^\infty(\mathbb{R}^n), \tag{3}$$

is equivalent to the isoperimetric inequality

$$(\mathrm{Vol}_n(\Omega))^{1/q} \leq C\mathrm{Vol}_{n-1}(\partial\Omega),$$

where $\partial\Omega$ is the boundary of a smooth bounded open set Ω. If we let $V(t)$ be the volume of a geodesic ball $B(t)$ of fixed centre and radius t, we have

$$\frac{\mathrm{d}}{\mathrm{d}t}\mathrm{Vol}(B(t)) = \mathrm{Vol}_{n-1}(\partial B(t)).$$

Hence, setting $\Omega = B(t)$ in the isoperimetric inequality, we get

$$\frac{\mathrm{d}}{\mathrm{d}t}V(t) \geq C^{-1}V(t)^{1/q}.$$

This shows that Sobolev's inequality (3) with $q = D/(D-1)$ implies $V(t) \geq ct^D$, which establishes a link between the Sobolev inequality and the volume growth function $V(t)$.

In 1958, E. Gagliardo and L. Nirenberg independently found the following elementary proof of the Sobolev inequality (3) in $\mathbb{R}^2$. For $f \in C_0^\infty(\mathbb{R}^2)$, $(x_0, y_0) \in \mathbb{R}^2$, note that

$$|f(x_0, y_0)| \leq \int_{-\infty}^{+\infty} |\frac{\partial f}{\partial x}(x, y_0)|\, dx, \int_{-\infty}^{+\infty} |\frac{\partial f}{\partial y}(x_0, y)|\, dy.$$

Therefore

$$|f(x_0, y_0)|^2 \leq \int_{-\infty}^{+\infty} |\frac{\partial f}{\partial x}(x, y_0)| \, dx \int_{-\infty}^{+\infty} |\frac{\partial f}{\partial y}(x_0, y)| \, dy.$$

Integrating over $\mathbb{R}^2$ with respect to x_0 and y_0, we get

$$||f||_2 \leq \left(||\frac{\partial f}{\partial x}||_1 ||\frac{\partial f}{\partial y}||_1 \right)^{1/2} \leq ||\frac{\partial f}{\partial x}||_1 + ||\frac{\partial f}{\partial y}||_1.$$

The same idea applies in $\mathbb{R}^n$, for $n \geq 3$, where one has to use Hölder's inequality. Finally, the L^p versions (1) all follow from the case $p = 1$. Indeed, applying (3) to f^s, where $f > 0$ and $s > 1$, we get

$$||f||_{sn/(n-1)}^s \leq Cs \int f^{s-1} |\nabla f| \, dx$$

$$\leq Cs \left(\int f^{(s-1)p'} \right)^{1/p'} \left(\int |\nabla f|^p \right)^{1/p},$$

where $\frac{1}{p'} + \frac{1}{p} = 1$. Choosing $s = p(n-1)/(n-p)$, one finds $(s-1)p' = np/(n-p)$ and thus

$$||f||_{np/(n-p)} \leq C \frac{p(n-1)}{n-p} ||\nabla f||_p.$$

When $p = 2$, $||\nabla f||_2^2 = (\Delta f, f) = ||\Delta^{1/2}||_2^2$. The above inequality becomes therefore

$$||f||_{2n/(n-2)} \leq C ||\Delta^{1/2}||_2, \ \forall f \in C_0^\infty(\mathbb{R}^n), \tag{4}$$

which is sometimes called a Dirichlet inequality; $(\Delta f, f)$ is the Dirichlet form associated with the heat diffusion semigroup $e^{-t\Delta}$. More generally, for $1 < p < +\infty$, $||\nabla f||_p \simeq ||\Delta^{\frac{1}{2}} f||_p, \ \forall f \in C_0^\infty(\mathbb{R}^n)$. This comes from the fact that the Riesz transforms $\frac{\partial}{\partial x_i} \Delta^{-1/2}$ (the higher dimensional analogues of the Hilbert transform), are bounded in $L^p(\mathbb{R}^n)$ when $1 < p < +\infty$. Together with (1), this yields

$$||f||_{np/(n-p)} \leq C ||\Delta^{1/2} f||_p, \quad \forall f \in C_0^\infty(\mathbb{R}^n), \tag{5}$$

for $1 < p < n$. However, there are ways to obtain (5) that avoid the use of (4). For instance, it can be shown that the operator $\Delta^{-\alpha/2}, 0 < \alpha p < n$, has a convolution kernel $k_\alpha(x) = c_\alpha |x|^{-n+\alpha}$. This kernel belongs to the weak L^p space $L^{n/(n-\alpha), \infty}$. Now, convolution with a function in $L^{r, \infty}$ sends L^p into L^q, for $1 + \frac{1}{q} = \frac{1}{p} + \frac{1}{r}$, provided that $1 < p < +\infty$, $1 < r < +\infty$ and $1 < \frac{1}{p} + \frac{1}{r}$. This yields

$$||f||_q \leq C ||\Delta^{\alpha/2} f||_p, \quad f \in C_0^\infty(\mathbb{R}^n),$$

for $1 < p < +\infty$, $0 < \alpha p < n$. It is worth emphasizing that the result is false for $p = 1$. In that case, only the weak inequality

$$||f||_{n/(n-\alpha), \infty} \leq C ||\Delta^{\alpha/2} f||_1, \quad f \in C_0^\infty(\mathbb{R}^n),$$

holds. Finally, we note that for $\alpha p > n$, Hölder continuity estimates hold instead.

I.2 Sobolev inequalities and the heat equation on Lie groups

Let us replace $\mathbb{R}^n$ with a connected unimodular Lie group G. Fix a set $\mathbf{X} = \{X_1, ..., X_k\}$ of left invariant vector fields. Without loss of generality, we may assume that $\mathbf{X}$ generates the Lie algebra of G. If it were not the case, we could always consider the sub-Lie algebra generated by $\mathbf{X}$ and work on the corresponding Lie subgroup of G. In this setting, it makes sense to ask for Sobolev inequalities of the type

$$\|f\|_q \leq C \sum_{i=1}^{k} \|X_i f\|_1, \ \forall f \in C_0^\infty(G). \tag{1}$$

If $\{X_1, ..., X_k\}$ is a *basis* of the Lie algebra, we can endow G with a left invariant Riemannian structure by deciding that this basis is orthonormal. Then $\sum_{i=1}^{k} \|X_i f\|_1 \simeq \|\nabla f\|_1$, where $\nabla f = (X_1 f, ..., X_k f)$ is the corresponding Riemannian gradient. Inequality (1) is then equivalent to an isoperimetric inequality and implies that the volume $V(t)$ of the geodesic balls of radius t satisfies $V(t) \geq ct^D$, $\forall t > 0$, where $q = D/(D-1)$.

One of our goals is to show that the reverse implication holds. In fact, even when $\mathbf{X}$ is not a basis but only generates the Lie algebra of G, there is a natural distance associated with $\mathbf{X}$. This distance, sometimes called the control distance, is defined by considering absolutely continuous paths that stay tangent almost everywhere to the fields $X_1, ..., X_k$. Let $V(t)$ be the volume of the balls $B(x, t)$ of radius t for that distance; of course $V(t)$ does not depend on the centre x. We will prove

Theorem   *The Sobolev inequality*

$$\|f\|_{D/(D-1)} \leq C \sum_{i=1}^{k} \|X_i f\|_1, \ \forall f \in C_0^\infty(G)$$

is equivalent to the volume growth condition

$$V(t) \geq ct^D, \ \forall t > 0.$$

To prove this theorem, our main tools will be the heat equation

$$\left(\frac{\partial}{\partial t} + \Delta \right) u = 0,$$

where $\Delta = \sum_{i=1}^{k} X_i^2$, and its fundamental solution h_t i.e. the heat kernel. The left invariance of the equation implies that h_t is a right convolution kernel, and we write $h_t(x, y) = h_t(y^{-1}x)$. In the case of $\mathbb{R}^n$, h_t is the Gauss kernel

$$(2\pi t)^{-n/2} e^{\frac{-|x-y|^2}{4t}}.$$

The property $h_{t+s} = h_t * h_s$ always holds and the heat semigroup H_t can be defined by

$$H_t f = f * h_t.$$

Equivalently, H_t is the Markov semigroup associated with the Dirichlet form $D(f, f) = \sum_{i=1}^k \|X_i f\|_2^2$.

The study of the properties of the solutions of the heat equation is of interest in itself, but one of the main themes of this book is the connection between these properties and Sobolev inequalities. The following argument, which borrows an idea of Nash, illustrates this point. Assume that the L^2 Sobolev (or Dirichlet) inequality,

$$\|f\|_{2n/(n-2)}^2 \leq C\|\nabla f\|_2^2 = (\Delta f, f), \tag{2}$$

holds. Then Hölder's inequality yields

$$\|f\|_2^{2+4/n} \leq C(\Delta f, f)\, \|f\|_1^{4/n}.$$

Setting $v(t) = \|h_t(x, .)\|_2^2$, we see that $v'(t) = -2(\Delta h_t, h_t)$ and, since $\|h_t\|_1 = 1$, the above yields the differential inequality $v(t)^{1+2/D} \leq -c\, v'(t)$. Integrating this, we get

$$v(t) \leq \left(\frac{2}{DC}t\right)^{-D/2}, \forall t > 0,$$

and therefore

$$\sup_{x \in G} h_t(x) = h_t(e) = \|h_{t/2}\|_2^2 \leq Ct^{-D/2}, \forall t > 0. \tag{3}$$

In fact, the hypothesis (2) and the conclusion (3) in the above argument are equivalent properties. This equivalence will be proved in the setting of abstract semigroups.

Theorem   *Let e^{-tA} be a symmetric submarkovian semigroup acting on L^2 of some measure space. For any $D > 2$, the following properties are equivalent :*

(i) $\|e^{-tA}f\|_\infty \leq t^{-D/2}\|f\|_1, \forall f \in L^1, \forall t > 0.$
(ii) $\|f\|_{2D/(D-2)} \leq C'\|A^{1/2}f\|_2, \forall f \in \mathcal{D}(A).$

This abstract result plays a central rôle in our approach.

Returning to the setting of Lie groups, we want to emphasize that the uniform estimate (3) can be complemented with bounds on $h_t(x)$ which include a Gaussian correction when x goes to infinity.

I.3 Harnack's principle

Let us say that Harnack's principle holds if there exists $C > 0$ such that for all $x \in G$ and $t > 0$, any positive solution u of $\left(\frac{\partial}{\partial t} + \Delta\right) u = 0$ in $]0, 4ts[\times B(x, 2\sqrt{s})$ satisfies

$$\sup_{B(x,\sqrt{s})} u(s,y) \leq C \inf_{B(x,\sqrt{s})} u(2s,y). \tag{1}$$

When it holds, the Harnack principle is a very powerful tool. To illustrate this, let us show how (1) yields bounds on the heat kernel in terms of the volume of balls.

Assume thus that Harnack's principle holds, and apply (1) to $u(t,x) = h_t(x)$, which is a solution in $]0, +\infty[\times G$. We conclude that there exists $C > 0$ such that

$$h_s(e) \leq C \inf_{B(e,\sqrt{s})} h_{2s}(x), \; \forall s > 0.$$

Integrating this inequality over the ball $B(e, \sqrt{s})$, we obtain

$$V(\sqrt{s}) h_s(e) \leq C \int_{B(e,\sqrt{s})} h_{2s}(x) \, dx \leq C$$

since $\|h_{2s}\|_1 = 1$. In particular, we see that $h_t(e) \leq Ct^{-D/2}$ as soon as Harnack's principle holds and $V(t) \geq ct^D$.

The drawback is that the Harnack principle does not always hold. Moreover, even if it does, it is not always easy to prove.

In our analytic and geometric study of Lie groups, each question has two different aspects. One corresponds to a local point of view, the other to a global one. The simplest instance of this is the behaviour of the volume growth function $V(t)$ which depends upon whether t tends to zero or to infinity. From a local point of view, the group structure does not play an important rôle, if any. Indeed, we will offer a local study of sums of squares of vector fields on manifolds. This will be based on dilation arguments and a local Harnack principle.

From the global point of view of geometry and analysis at infinity, the group structure enters into play in an essential manner. Indeed, in the setting of Lie groups, our basic result is that the Sobolev inequality

$$\|f\|_{D/(D-1)} \leq C \|\nabla f\|_1, \; \forall f \in C_0^\infty(G),$$

the kernel estimate

$$h_t(e) \leq Ct^{-D/2}, \; \forall t > 0,$$

and the volume growth condition

$$V(t) \geq ct^D, \; \forall t > 0$$

are equivalent. This equivalence simply fails to hold if one replaces the group G by – say – a Riemannian manifold with bounded geometry.

I.4 A guide to this book

In Chapter II, we build the semigroup machinery that will enable us to link the Sobolev inequalities with the behaviour of the heat kernel. These functional analytic results are of independent interest.

In Chapter III we describe some basic properties of the sums of squares of vector fields. A given set of vector fields $X_1, ..., X_{k+1}$ satisfies the Hörmander condition if the fields $X_1, ..., X_{k+1}$ together with their brackets of every order span the tangent space at each point. Under this condition, a genuine distance can be defined by considering the "minimal length" of absolutely continuous paths tangent to the fields $X_1, ..., X_{k+1}$. Moreover, the operator $\sum_{i=1}^{k} X_i^2 + X_{k+1}$ is hypoelliptic (Hörmander's theorem) and a local Harnack inequality holds.

Chapter IV focuses on the study of the sublaplacian associated with a Hörmander system of left invariant vector fields on a nilpotent Lie group. Here our analysis is based on Harnack's principle. Indeed, any connected nilpotent Lie group can be covered by another nilpotent Lie group that admits a dilation structure. This dilation structure, together with the local Harnack principle derived in Chapter III, yields the scaled Harnack principle described in Section 3 above. This principle tranfers easily to G. From this, a two-sided Gaussian bound for the heat kernel follows. We also study in detail the volume growth of nilpotent Lie groups. This shows the existence of a local dimension d that governs the behaviour of the volume of small balls, and of a dimension at infinity D that governs the volume of large balls. Finally, heat kernel and volume estimates, together with Chapter II, yield optimal Sobolev inequalities.

Chapter IV also serves as a model for a general study of Hörmander systems of vector fields. In Chapter V, we show how Harnack's principle and a local scaling technique yield satisfactory local results for the heat equation associated with sublaplacians on groups and manifolds.

Chapter VI introduces in the simple setting of discrete groups the main ideas leading to the analytic and geometric study of groups at infinity. In order to stay away from technicalities, the results are not stated in their optimal form. However, they are more than enough to show that the only recurrent finitely generated groups are the finite extensions of $\{0\}, \mathbb{Z}$, and $\mathbb{Z}^2$.

Chapter VII develops the various tools needed to extend and refine the results of Chapter VI. The main result establishes the sharp relationship between volume growth and decay of convolution powers, in the setting of locally compact, compactly generated groups. In the process, we give an analogue of the theory of Chapter II for discrete time semigroups.

Chapter VIII considers unimodular connected Lie groups. Here, the functional analytic tools of Chapter II play an essential part. Together with the local results of Chapter V, they yield, in the case of polynomial volume growth, two-sided Gaussian estimates for the heat kernel, optimal Sobolev inequalities, and Harnack's principle. In the case of exponential volume growth, we prove a sharp result concerning the uniform decay of the heat kernel at infinity. Chapter IX gives up the study of the heat kernel and concentrates on Sobolev inequalities for non-unimodular Lie groups. An inequality of Hardy, ideas from Chapter VII and the splitting of G as the semi-direct product $G \simeq \overline{G} \rtimes R$, where $\overline{G}$ is the kernel of the modular function, are the essential ingredients.

Finally Chapter X contains various geometric applications of the above theory.

This book does not aim at being self-contained. A background in functional analysis, differential geometry and Lie group theory would certainly be helpful to the reader. However, not much is needed to understand our geometric and analytic study of discrete groups in Chapter VI. Also it is *not* necessary to master the theory of one-parameter semigroups of operators to follow Chapter II, and it is certainly not necessary to master the details of Lie group theory to make one's way through Chapters IV, VIII, and IX. The only important result whose proof is not given, but which is nevertheless fundamental (for the local theory) is Hörmander's theorem.

All references, including the bibliography for the background material we use in the text, are to be found in the References and Comments section at the end of each chapter.

CHAPTER II

DIMENSIONAL INEQUALITIES FOR
SEMIGROUPS OF OPERATORS ON THE L^p SPACES

II.1 Introduction, notation

Let $T_t = e^{-tA}$ be a symmetric submarkovian semigroup on a measure space (see Section 5 for definitions). The main theme of this chapter is the equivalence between the following two properties, for $n \in]2, +\infty[$:

$$D_n : \quad \exists C \text{ such that } \|f\|_{2n/(n-2)} \le C(Af, f)^{\frac{1}{2}}, \quad \forall f \in \mathcal{D}(A),$$

$$R_n : \quad \exists C \text{ such that } \|T_t f\|_\infty \le C t^{-n/2} \|f\|_1, \quad \forall f \in L^1, \forall t > 0.$$

A number n, if there is any, such that D_n or R_n is fulfilled may be called the dimension of the semigroup e^{-tA}: for example, if n is an integer greater than 3, the heat semigroup on $\mathbb{R}^n$, $e^{-t\Delta}$, has dimension n and the Poisson semigroup, $e^{-t\Delta^{\frac{1}{2}}}$, has dimension $2n$. This justifies the title of this chapter. Nevertheless, a semigroup may have no dimension (e.g. the trivial semigroup) or several ones: if the measure space under consideration is discrete, a semigroup of dimension n is also of dimension m, for every $m \le n$, since the ℓ^p spaces are nested.

The implication "$R_n \Rightarrow D_n$" relies on an abstract analogue of the classical Hardy-Littlewood-Sobolev theory which will be developed in Section 2. The implication "$D_n \Rightarrow R_n$" can be obtained in several ways. We shall present three of them in Section 3. The first one uses the analyticity of the semigroup. The other two are inspired by ideas of Nash and Moser. None of these methods are limited to the framework of symmetric submarkovian semigroups, and this is important for the applications. Nevertheless, this setting is central, and we shall consider it specifically in Section 5. In Section 4, properties D_n and R_n are localized for small and large time.

From now on, (X, ξ) will be a measure space and T_t, $t \ge 0$, a semigroup of operators defined on $L^1 \cap L^\infty$ which, for every $p \in [1, +\infty]$, extends to a semigroup on L^p, of class C^0 if $p \ne +\infty$ (i.e. $T_t f \to f$ in L^p when $t \to 0^+$). Define A_p by $A_p f = \lim_{t \to 0^+} \frac{f - T_t f}{t}$, when this limit exists in L^p; $-A_p$ is the infinitesimal generator of T_t on L^p. Denote by $\mathcal{D}$ a vector space dense in L^p, and dense for the graph norm in the domain of A_p, for every finite p. For instance, it is easy to see that the space

$$\mathcal{D}_0 = \text{Vect}\left\{ \int_0^{+\infty} \phi(t) T_t f \, dt \mid \phi \in C_0^\infty(]0, +\infty[), f \in L^\infty, \xi(\{f \ne 0\}) < +\infty \right\}$$

always fulfills these conditions. We shall thus omit the index p and set $A = A_p$, $p \in [1, +\infty[$.

A harmonic function with respect to T_t is a function $v: \mathbb{R}^{+*} \times X \to \mathbb{C}$ such that

$$\forall t > 0, \quad v_t(.) = v(t, .) \in L^1 + L^\infty$$

$$\forall t, s > 0, \quad T_t v_s = v_{t+s}.$$

For example, if $f \in L^1 \cap L^\infty$, $v(t, x) = T_t f(x)$ is harmonic with respect to T_t. Given a harmonic function v, denote by v^* the maximal function defined by $v^*(x) = \sup_{t>0} |v(t, (x))|$. For $p \in]0, +\infty[$, we shall say that a function v which is harmonic with respect to T_t belongs to H^p if v^* belongs to L^p. The quasi-norm $\|v\|_{H^p} = \|v^*\|_p$ is a norm when $p \geq 1$.

II.2 Hardy–Littlewood–Sobolev theory

The property R_n is a regularization property which may be generalized for $0 < p < q \leq +\infty$ as follows:

$$R(n, p, q) \quad \|v(t, \cdot)\|_q \leq C t^{-n(\frac{1}{p} - \frac{1}{q})/2} \sup_{s>0} \|v(s, \cdot)\|_p, \quad \forall t > 0, \forall v \text{ harmonic.}$$

II.2.1 Proposition *Suppose there exist $0 < p < q \leq +\infty$ and $n > 0$ such that*

$$\|T_t f\|_q \leq C t^{-n(\frac{1}{p} - \frac{1}{q})/2} \|f\|_p, \quad \forall t > 0, \forall f \in L^p.$$

Then $R(n, p_1, q)$ is satisfied for every $p_1 \leq p$.

Proof Using the hypothesis and Hölder's inequality, one has, putting $\alpha = n(\frac{1}{p} - \frac{1}{q})/2$ and $\frac{1}{p} = \frac{\theta}{q} + \frac{1-\theta}{p_1}$,

$$\|T_{2t} f\|_q \leq C t^{-\alpha} \|T_t f\|_p \leq C t^{-\alpha} \|T_t f\|_q^\theta \|T_t f\|_{p_1}^{1-\theta}.$$

Set

$$K(f, r) = \sup_{t \in [0,r]} \{ t^{\alpha/(1-\theta)} \|T_t f\|_q (\sup_{s>0} \|T_s f\|_{p_1})^{-1} \}.$$

One has

$$\|T_{2t} f\|_q \leq C K^\theta(f, r) t^{-\alpha/(1-\theta)} \sup_{s>0} \|T_s f\|_{p_1}, \quad t \in]0, 1].$$

Thus

$$K(f, r) \leq C K^\theta(f, r).$$

Therefore $K(f, r) \leq C^{1/(1-\theta)}$ for f such that $\sup_{s>0} \|T_s f\|_{p_1} < +\infty$. The property $R(n, p_1, q)$ easily follows.

If T_t is equicontinuous on L^1 and L^∞, $R(n, p, q)$ may be reformulated, for $1 \leq p < q \leq +\infty$, as

$$\|T_t\|_{p \to q} \leq C t^{-n(\frac{1}{p} - \frac{1}{q})/2}, \quad \forall t > 0.$$

Thus one has

II.2.2 Proposition *Let T_t be a semigroup which is equicontinuous on L^1 and L^∞, and let $n > 0$. Then:*

(i) *R_n is equivalent to each $R(n, p, q)$, for $1 \leq p < q \leq +\infty$.*

(ii) R_n implies $R(n, p, q)$ for all p, q such that $0 < p < q < +\infty$.

Proof If $0 < p < q \leq +\infty$, Hölder's inequality

$$\|v(t, \cdot)\|_q \leq \|v(t, \cdot)\|_\infty^{1-p/q} \|v(t, \cdot)\|_p^{p/q},$$

shows that $R(n, p, \infty)$ always implies $R(n, p, q)$. Moreover, the Riesz–Thorin theorem shows that $\|T_t\|_{\infty \to \infty} \leq C_1$ and $\|T_t\|_{p \to \infty} \leq C_2 t^{-n/2p}$ imply

$$\|T_t\|_{r \to \infty} \leq C t^{-n/2r} \quad \text{for } 1 \leq p \leq r \leq +\infty.$$

It follows that R_n implies $R(n, p, q)$ for $1 \leq p < q \leq +\infty$ if T_t is equicontinuous on L^∞; Proposition II.2.1 then easily gives (ii). To end the proof of (i), fix $1 \leq p < q \leq +\infty$ and suppose that $R(n, p, q)$ is satisfied. Proposition II.2.1 then shows that $R(n, 1, q)$ is satisfied, which means, since T_t is equicontinuous on L^1, that

$$\|T_t f\|_q \leq C t^{-n/2q'} \|f\|_1, \quad \forall t > 0, \quad \forall f \in L^1,$$

where $\frac{1}{q} + \frac{1}{q'} = 1$. The dual semigroup T_t^* thus satisfies

$$\|T_t^* f\|_\infty \leq C t^{-n/2q'} \|f\|_{q'}, \quad \forall t > 0, \quad \forall f \in L^{q'}.$$

Applying Proposition II.2.1 to T_t^*, one gets

$$\|T_t^* f\|_\infty \leq C t^{-n/2} \|f\|_1, \quad \forall t > 0, \quad \forall f \in L^1.$$

By duality, one obtains the same estimate for T_t, i.e. T_t satisfies R_n.

We are now going to see that R_n implies mapping properties for the potential operators $G_\zeta f = \Gamma(\zeta/2) A^{-\zeta/2} f$, for $\zeta \in \mathbb{C}$, $\text{Re } \zeta > 0$. One may take as a definition of these operators

$$G_\zeta f = \int_0^{+\infty} t^{(\zeta/2)-1} T_t f \, dt$$

when the integral converges at $+\infty$. Note that if v is harmonic and if $G_\zeta v$ exists, $G_\zeta v$ is also a harmonic function.

II.2.3 Proposition   *Let $n > 0$ and $0 < p < +\infty$. Suppose that T_t satisfies $R(n, p, \infty)$, and let $q > 0$ and $\zeta \in \mathbb{C}$, $\text{Re } \zeta = \gamma > 0$, be such that $\frac{1}{q} = \frac{1}{p} - \frac{\gamma}{n}$. Then, for every harmonic function v belonging to H^p, $G_\zeta v$ exists and*

$$(G_\zeta v)^*(x) \leq C v^*(x)^{p/q} \|v^*\|_p^{1-(p/q)}.$$

In particular, G_ζ is bounded from H^p to H^q.

Proof   Let us write

$$G_\zeta v(s, x) = \int_0^T t^{(\zeta/2)-1} v(t+s, x) \, dt + \int_T^{+\infty} t^{(\zeta/2)-1} v(t+s, x) \, dt.$$

Using $R(n, p, +\infty)$ to estimate the second integral, we get

$$|G_\zeta v(s, x)| \leq 2\gamma^{-1}T^{\gamma/2}v^*(x) + CT^{\gamma/2-n/2p}||v^*||_p.$$

Choosing $T = (v^*(x)/||v^*||_p)^{-2p/n}$, we obtain, since $\frac{1}{q} = \frac{1}{p} - \frac{\gamma}{n}$,

$$(G_\zeta v)^*(x) \leq Cv^*(x)^{p/q}||v^*||_p^{1-p/q},$$

hence

$$||(G_\zeta v)^*||_q \leq C||v^*||_p.$$

Propositions II.2.3 and II.2.2 yield finally

II.2.4 Theorem *If T_t is equicontinuous on L^1 and L^∞, and satisfies R_n for some $n > 0$, then for every $0 < p < +\infty$ and $\zeta \in \mathbb{C}$, $\operatorname{Re}\zeta = \gamma > 0$, G_ζ is well defined on H^p and bounded from H^p to H^q for q such that $\frac{1}{q} = \frac{1}{p} - \frac{\gamma}{n} > 0$.*

In the case where T_t is symmetric submarkovian, the norms $||v^*||_p$ and $||f||_p$, where $v(t,.) = T_t f(.)$, are equivalent for $p > 1$ and Theorem II.2.4 gives the behaviour of G_ζ on L^p for $p > 1$. More general results can be obtained by using the weak type spaces. For $p \geq 1$, denote by $L^{p,\infty}$ the space of functions f such that

$$\sup_{\lambda>0} \lambda^p \xi(\{x \in X \mid |f(x)| > \lambda\}) = ||f||_{p,\infty} < +\infty.$$

An operator T which is bounded from L^p to $L^{q,\infty}$, i.e. which satisfies

$$||Tf||_{q,\infty} \leq C||f||_p, \quad \forall f \in L^p,$$

is said to be of weak type (p, q); the classical notation $L^{p,\infty}$ refers to the scale of Lorentz spaces $L^{p,q}$. Recall now the well-known interpolation theorem of Marcinkiewicz.

II.2.5 Theorem *Let T be a (p_i, q_i), $i = 1, 2$ weak type operator with $1 \leq p_i \leq q_i \leq +\infty$, and $q_1 \neq q_2$. Then T is bounded from L^p to L^q for all p, q such that $\frac{1}{p} = \frac{\theta}{p_1} + \frac{(1-\theta)}{p_2}$, $\frac{1}{q} = \frac{\theta}{q_1} + \frac{(1-\theta)}{q_2}$, $0 < \theta < 1$.*

Let us come back to the potential operators G_ζ; one has

II.2.6 Proposition *Let $n > 0$, $1 \leq p < +\infty$. Suppose that T_t is equicontinuous on L^p and satisfies $R(n, p, \infty)$. Let $\zeta \in \mathbb{C}$, $\operatorname{Re}\zeta = \gamma > 0$, and q such that $\frac{1}{q} = \frac{1}{p} - \frac{\gamma}{n} > 0$. Then $G_\zeta f$ exists for every $f \in L^p$ and G_ζ is of weak type (p, q).*

Proof Suppose that $||f||_p = 1$ and write again

$$G_\zeta f(x) = F^T(x) + F_T(x), \quad \text{where } F^T(x) = \int_0^T t^{(\zeta/2)-1}T_t f(x) \, dt.$$

Then

$$\xi\left(\{x \mid |G_\zeta f(x)| > \lambda\}\right) \le \xi\left(\{x \mid |F^T(x)| > \lambda/2\}\right) + \xi\left(\{x \mid |F_T(x)| > \lambda/2\}\right).$$

Now, applying $R(n, p, \infty)$, one sees that

$$\|F_T\|_\infty \le CT^{\gamma/2 - n/2p}.$$

Choose T_0 such that $CT_0^{-n/2q} = CT_0^{\gamma/2-n/2p} = \lambda/4$. The Bienaymé-Tchebychev inequality yields

$$\xi\left(\{x \mid |G_\zeta f(x)| > \lambda\}\right) \le \xi\left(\{x \mid |F^{T_0}(x)| > \lambda/2\}\right) \le C\left(\|F^{T_0}\|_p/\lambda\right)^p.$$

Now

$$\|F^{T_0}\|_p = \|\int_0^{T_0} t^{(\gamma/2)-1}T_t f\, dt\|_p \le CT_0^{\gamma/2}.$$

Finally

$$\xi\left(\{x \mid |G_\zeta f(x)| > \lambda\}\right) \le CT_0^{p\gamma/2}\lambda^{-p} = C'\lambda^{-q}, \quad \lambda > 0,$$

which is the desired conclusion.

II.2.7 Theorem *Let T_t be a semigroup which is equicontinuous on L^1 and L^∞, and satisfies R_n, for $n > 0$. Then, for $\zeta \in \mathbb{C}$, $\mathrm{Re}\,\zeta = \gamma > 0$, we have:*
(i) *For $\gamma < n$, G_ζ is of weak type $(1, n/(n-\gamma))$.*
(ii) *If $1 < p < +\infty$, and $\frac{1}{q} = \frac{1}{p} - \frac{\gamma}{n} > 0$, G_ζ is bounded from L^p to L^q.*

Proof It suffices to use Propositions II.2.2 and II.2.6 and the Marcinkiewicz theorem II.2.5.

II.2.8 Remark Proposition II.2.6 and Theorem II.2.7 still hold if one replaces the operator G_ξ by $\tilde{G} = \int_0^{+\infty} S_t\, dt$, where S_t is any family of operators satisfying $\|S_t\|_{p\to p} \le Mt^{(\gamma/2)-1}$ and $\|S_t\|_{1\to\infty} \le Mt^{(\gamma/2)-1-n/2}$, for $t > 0$ and $1 \le p \le +\infty$.

II.3 Converses to the Hardy–Littlewood–Sobolev theory

The aim of this section is to show that the converse of Theorem II.2.7 is valid. The equivalence between D_n type and R_n type properties will then be the powerhouse of the theory developed in this book.

Let us recall some classical facts concerning fractional powers of operators, subordination and analytic semigroups. These facts are explained in several textbooks; we quote some of them in the References and Comments at the end of this chapter. Given an equicontinuous semigroup T_t of class C^0 on a Banach space X, the *subordinated semigroup* of order $\alpha \in]0, 1[$ of T_t is the semigroup

$$T_t^\alpha = \int_0^{+\infty} f_{t,\alpha}(v)T_v\, dv,$$

where $f_{t,\alpha}$ has for Laplace transform the function $\lambda \to e^{-t\lambda^\alpha}$ and hence satisfies

$$\int_0^{+\infty} f_{t,\alpha}(\lambda)\, d\lambda = 1 \quad \text{and} \quad f_{t,\alpha} \geq 0.$$

It immediately follows that T_t^α is a semigroup of class C^0 on X. From the above, one also deduces

$$T_t^\alpha = \int_0^{+\infty} f_{1,\alpha}(v) T_{vt^{1/\alpha}}\, dv.$$

Using classical Tauberian theorems for Laplace transforms, it is easy to estimate $f_{1,\alpha}(v)$ near 0, and to get

$$\int_0^{+\infty} v^{-n/2} f_{1,\alpha}(v)\, dv < +\infty, \quad \forall n > 0.$$

This shows that, if T_t acts on L^p and satisfies R_n, then T_t^α satisfies $R_{n/\alpha}$.

Let $-A$ (resp. $-A^\alpha$) be the infinitesimal generator of T_t (resp. T_t^α). We then have, on the domain of A,

$$A^\alpha = \Gamma(-\alpha)^{-1} \int_0^{+\infty} s^{-\alpha-1}(T_s - I)\, ds = \frac{\sin \alpha \pi}{\pi} \int_0^{+\infty} s^{\alpha-1}(sI + A)^{-1}\, ds$$

and

$$A^\alpha A^\beta = A^{\alpha+\beta}, \quad 0 < \alpha, \beta, \alpha + \beta < 1,$$
$$(A^\alpha)^\beta = A^{\alpha\beta}, \quad 0 < \alpha, \beta < 1.$$

Those formulas remain valid for $\alpha, \beta \in \mathbb{R}_+$, when restricted to suitable domains.

Recall that T_t is said to be analytic if it admits an extension T_ζ in a sector $\{\zeta \in \mathbb{C} \mid \operatorname{Arg}\zeta < \theta\}$, $\theta > 0$. It is said to be bounded analytic if it admits an equicontinuous extension to a sector $\{\zeta \in \mathbb{C} \mid \operatorname{Arg}\zeta \leq \theta\}$, $\theta > 0$. The former property is equivalent to the estimate

$$\|AT_t\|_{X \to X} \leq Ct^{-1}, \quad \forall t \in]0, 1[,$$

and the latter to

$$\|AT_t\|_{X \to X} \leq Ct^{-1}, \quad \forall t > 0.$$

We then have, for every $\alpha > 0$, $\forall t \in]0, 1[$, resp. $\forall t > 0$:

$$\|A^\alpha T_t\|_{X \to X} \leq Ct^{-\alpha}.$$

One can show that a subordinate semigroup is always bounded analytic. The same holds for a semigroup of self-adjoint contractions on L^2. In this case, if the semigroup is, in addition, equicontinuous on L^1, it is bounded analytic on L^p, $1 < p < +\infty$. The L^2 result easily follows from spectral theory and the L^p result, in a more intricate way, from complex interpolation.

Let us finally recall that if one puts, for $\alpha > 0$ and T_t bounded analytic,

$$A^{-\alpha} = \Gamma(\alpha)^{-1} \int_0^{+\infty} t^{\alpha-1} T_t \, dt = \Gamma(\alpha)^{-1} G_\alpha,$$

one has, under the hypothesis that A is one-to-one on $\mathcal{D}(A)$,

$$A^{-\alpha} A^\alpha = \mathrm{Id} \quad \text{on } \mathcal{D}(A) \text{ for } 0 < \alpha < 1.$$

Again, the restriction on α can be removed if one considers a smaller domain.

II.3.1 Theorem *Suppose that T_t is equicontinuous on L^1 and L^∞, and that there exist $\alpha > 0$, $1 < p < q \le +\infty$ such that*

$$\|f\|_q \le C \|A^{\alpha/2} f\|_p, \quad \forall f \in \mathcal{D}.$$

Suppose moreover that T_t is bounded analytic on L^p. Then T_t satisfies R_n, where n is given by $\frac{1}{q} = \frac{1}{p} - \frac{\alpha}{n}$.

Proof By the hypothesis, $A^{-\alpha/2}$ is a bounded operator from L^p to L^q. Let us write $T_t = A^{-\alpha/2} A^{\alpha/2} T_t$, hence

$$\|T_t\|_{p \to q} \le \|A^{-\alpha/2}\|_{p \to q} \|A^{\alpha/2} T_t\|_{p \to p} \le C t^{-\alpha/2},$$

where the last inequality follows from the analyticity of T_t. It suffices now to apply Proposition II.2.1.

We shall now present two other methods leading to R_n, which do not involve any analyticity or symmetry assumption. The first one is inspired by an idea of Nash.

II.3.2 Theorem *Suppose that T_t is equicontinuous on L^1. Let $n > 0$. Suppose there exists $C > 0$ such that*

$$\|f\|_2^{2+4/n} \le C \mathrm{Re} \, (Af, f) \|f\|_1^{4/n}, \quad \forall f \in \mathcal{D}.$$

Then there exists K such that $\|T_t\|_{1 \to 2} \le K t^{-n/4}$ for all $t > 0$. If moreover T_t is equicontinuous on L^∞, it satisfies

$$\|T_t\|_{1 \to \infty} \le K t^{-n/2}, \quad \forall t > 0.$$

Proof Fix $f \in \mathcal{D}$ with $\|f\|_1 = 1$. Since

$$\frac{d}{dt} \|T_t f\|_2^2 = -2\mathrm{Re} \, (AT_t f, T_t f),$$

the hypothesis yields

$$\frac{d}{dt} \|T_t f\|_2^2 \le -2C^{-1} \|T_t f\|_2^{2+4/n} \|T_t f\|_1^{-4/n} \le -2C^{-1} M^{-4n} \|T_t f\|_2^{2+4/n}$$

i.e.

$$\frac{d}{dt}\left(\|T_tf\|_2^{-4/n}\right) \geq 4n^{-1}C^{-1}M^{-4/n}, \quad \forall t > 0.$$

This differential inequality yields

$$\|T_tf\|_2 \leq M(nC)^{n/4}t^{-n/4}, \quad \forall t > 0.$$

The second assertion immediately follows by duality (or by II.2.2).

II.3.3 Remarks

(a) If A is self-adjoint on L^2, there is a converse to II.3.2; suppose indeed that

$$\|T_t\|_{1\to 2} \leq Kt^{-n/4}, \quad \forall t > 0.$$

Write

$$\|f\|_2^2 = \|T_tf\|_2^2 - \int_0^t \frac{d}{ds}\|T_sf\|_2^2\, ds \leq Kt^{-n/4}\|f\|_1 + \int_0^t (AT_sf, T_sf)\, ds$$
$$\leq Kt^{-n/4}\|f\|_1 + t(Af, f).$$

The last inequality follows from the fact that $s \mapsto (AT_sf, T_sf) = \|A^{1/2}T_sf\|_2^2$ is a non-increasing function. One concludes by optimizing on $t > 0$.

(b) The proof of II.3.1 can be adapted to show that, if $A^{-\alpha/2}$ is of weak type (p, q) for some α, p, q such that $p > 1$, $0 < \alpha p < n$ and $\frac{1}{q} = \frac{1}{p} - \frac{\alpha}{n}$, then T_t satisfies R_n; in II.3.2 also, one can replace the hypothesis by a weak type hypothesis.

(c) One may also characterize R_n in terms of the resolvent instead of the generator. For instance, one can show: T_t satisfies R_n if and only if $\|\lambda J_\lambda\|_{p\to q} \leq C\lambda^{\alpha/2}$, for small enough $\alpha = n(\frac{1}{p} - \frac{1}{q}) > 0$, and $J_\lambda = (\lambda I + A)^{-1}$.

We are now going to state a result of the same kind, whose proof stems from the work of J. Moser. The iterative process of Moser that we are going to describe in an abstract form has the following formula as a starting point:

$$\frac{d}{dt}\|T_tf\|_p^p = -p\,\mathrm{Re}\,(AT_tf, (T_tf)_p), \quad \forall f \in \mathcal{D},$$

where $p > 1$, $f_p = \mathrm{sgn}(f)|f|^{p-1}$ and $\mathrm{sgn}(\zeta) = 0$ if $\zeta = 0$, $\mathrm{sgn}(\zeta) = \zeta/|\zeta|$ otherwise. This formula is a straightforward consequence of the formula for the differential of the L^p norm.

II.3.4 Theorem   *Let $n \geq 2$, and T_t equicontinuous on L^1 and L^∞, such that*

$$\|f\|_{pn/(n-2)}^p \leq C\mathrm{Re}\,(Af, f_p), \quad \forall f \in \mathcal{D},$$

for some $p \in]1, +\infty[$. Then

$$\|T_t\|_{1\to\infty} \leq Ct^{-n/2}, \quad \forall t > 0.$$

Proof Let $f \in \mathcal{D}$. We have

$$||f||_p^p \geq ||f||_p^p - ||T_t f||_p^p = \int_0^t -\frac{d}{ds}||T_s f||_p^p\, ds = p \int_0^t \mathrm{Re}\,(AT_s f, (T_s f)_p)\, ds.$$

Set $q = \frac{np}{n-2}$. By the hypothesis,

$$||f||_p^p \geq \frac{1}{C}\int_0^t ||T_s f||_q^p\, ds.$$

Now, since T_t is equicontinuous on L^q,

$$t||T_t f||_q^p \leq C\int_0^t ||T_s f||_q^p\, ds,$$

and we get

$$||T_t f||_q \leq Ct^{-1/p}||f||_p.$$

Proposition II.2.1 again ends the proof.

Let us make a general comment about the equicontinuity assumption. If T_t is simply strongly continuous on L^1 and L^∞, it follows from the semi-group property that there exist $M \geq 1$ and $\omega \in \mathbb{R}$ such that $||T_t||_{p \to p} \leq Me^{\omega t}$, $\forall t \geq 0$. In other words, there exists ω such that the semigroup $e^{\omega t}T_t$ is equicontinuous on L^1 and L^∞. One can thus reformulate Theorems II.3.2 and II.3.4 for general semigroups acting on L^1 and L^∞. However, such statements are of limited value: it is not easy in general to get meaningful information about ω.

Here is a more sophisticated version of Theorem II.3.4:

II.3.5 Theorem   *Let $n > 2$. Suppose there exist C_p and γ_p, some positive functions of p majorized by a polynomial, with $\gamma_2 = 1$, such that*

$$\forall p \geq 2, \forall f \in \mathcal{D}, \quad ||f||_{pn/(n-2)}^p \leq C_p \mathrm{Re}\,((A + \alpha \gamma_p)f, f_p). \tag{H}$$

Then there exists K which only depends on n, C_p and γ_p such that

$$||T_t||_{2 \to \infty} \leq K(1 + \alpha t)^{n/4}t^{-n/4}e^{\alpha t}, \quad \forall t > 0.$$

Proof We have to show that

$$||T_t||_{2 \to \infty} \leq Kt^{-n/4}, \forall \alpha, t > 0 \text{ such that } \alpha t \leq 1,$$

and that

$$||T_t||_{2 \to \infty} \leq K\alpha^{n/4}e^{\alpha t}, \forall \alpha, t > 0 \text{ such that } \alpha t \geq 1.$$

But the second estimate easily follows from the first one and from the fact that $\mathrm{Re}\,((A+\alpha)f, f) \geq 0, \forall f \in \mathcal{D}$, which in turn follows from (H) with $p = 2$.

Indeed the first estimate yields in particular $||T_{1/\alpha}||_{2\to\infty} \le K\alpha^{n/4}, \forall \alpha > 0$.
Now $Re((A+\alpha)f, f) \ge 0, \forall f \in \mathcal{D}$ implies $||T_t||_{2\to 2} \le e^{\alpha t}, \forall \alpha, t > 0$. Finally,
for $\alpha t \ge 1$,

$$||T_t||_{2\to\infty} \le ||T_{1/\alpha}||_{2\to\infty}||T_{t-\frac{1}{\alpha}}||_{2\to 2} \le K\alpha^{n/4}e^{\alpha(t-\frac{1}{\alpha})}.$$

Let us now prove the first estimate. Fix $p \ge 2$ and $q = \frac{np}{n-2}$; let $\tilde{T}_t = e^{-\alpha\gamma_p t}T_t$. The hypothesis (H) means that $||\tilde{T}_t f||_q^p \le -C_p\frac{d}{dt}||\tilde{T}_t f||_p^p$. One
gets by integration

$$\int_0^t e^{-\alpha p\gamma_p s}||T_s f||_q^p ds \le C_p(||f||_p^p - e^{-\alpha p\gamma_p t}||T_t f||_p^p), \forall\, t > 0, \forall\, f \in \mathcal{D}.$$

This implies, since one can always assume γ_p non-decreasing,

$$\int_0^t e^{-\alpha p\gamma_q s}||T_s f||_q^p ds \le C_p||f||_p^p, \forall\, t > 0, \forall\, f \in \mathcal{D}.$$

Now $Re((A+\alpha\gamma_q)f, f_q)$ is positive, hence $e^{-\alpha\gamma_q t}T_t$ is a contraction semigroup
on $L^q(X, \xi)$. Therefore $e^{-\alpha p\gamma_q t}||T_t f||_q^p \le e^{-\alpha p\gamma_q s}||T_s f||_q^p$, for $s < t$. This
yields

$$||T_t||_{p\to q} \le C_p^{1/p}e^{\alpha\gamma_q t}t^{-1/p}, \forall\, t > 0.$$

Set now $k = \frac{n}{n-2}$, $p_\nu = 2k^\nu$ for $\nu \in \mathbb{N}$, and observe that $\sum_{\nu=0}^{+\infty} 1/p_\nu = n/4$.
For every sequence (t_ν) of positive real numbers such that $\sum_{\nu=0}^{+\infty} t_\nu = 1$, one
has

$$||T_t||_{2\to+\infty} \le \Pi_{\nu=0}^{+\infty}||T_{tt_\nu}||_{p_\nu\to p_{\nu+1}}.$$

Together with the above estimate, this gives

$$||T_t||_{2\to\infty} \le \Pi_{\nu=0}^{+\infty}C_{p_\nu}^{1/p_\nu}\Pi_{\nu=0}^{+\infty}t_\nu^{-1/p_\nu}e^{\alpha t\sum_{\nu=0}^{+\infty}\gamma_{p_{\nu+1}}t_\nu}t^{-n/4}, \forall Jt > 0.$$

Since the growth of C_p is polynomial, one has $\Pi_{\nu=0}^{+\infty}C_{p_\nu}^{1/p_\nu} < +\infty$. Since
the growth of γ_p is polynomial as well, one can choose the t_ν so that
$\Pi_{\nu=0}^{+\infty}t_\nu^{-1/p_\nu} < +\infty$ and $\sum_{\nu=0}^{+\infty}\gamma_{p_{\nu+1}}t_\nu < +\infty$. This ends the proof of the
proposition.

II.4 Localizations

A consequence of Theorems II.2.7 and II.3.1 is the following equivalence:

II.4.1 Theorem *Suppose that T_t is equicontinuous on L^1 and L^∞, and
bounded analytic on L^2. The following properties are equivalent, for $n > 0$:*

(i) $||f||_{2n/(n-2\alpha)} \le C_\alpha||A^{\alpha/2}f||_2$, *for all $f \in \mathcal{D}$, for one or all $\alpha \in]0, n/2[$:*

(ii) T_t *satisfies R_n.*

We are now going to characterize in a similar way the properties:

$$R_n(0): \quad ||T_t||_{1\to\infty} \le Ct^{-n/2}, \quad \forall t \in]0, 1[.$$
$$R_n(\infty): \quad ||T_t||_{1\to\infty} \le Ct^{-n/2}, \quad \forall t \ge 1.$$

By dealing with the semigroup $\widetilde{T_t} = e^{-t}T_t$, one reduces the study of $R_n(0)$ to that of R_n. More precisely, we can state

II.4.2 Theorem   *Suppose that T_t is equicontinuous on L^1 and L^∞, and analytic on L^2. Let n and α be such that $0 < \alpha < n/2$; then the following properties are equivalent:*

(i) $||f||_{2n/(n-2\alpha)} \leq C(||A^{\alpha/2}f||_2 + ||f||_2)$, *for all $f \in \mathcal{D}$;*

(ii) $||(I - T_1)^k f||_{2n/(n-2\alpha)} \leq C||A^{\alpha/2}f||_2$, *for all $f \in \mathcal{D}$, where $k > \alpha/2$, and $||T_1||_{2\to 2n/(n-2\alpha)} < +\infty$;*

(iii) *T_t satisfies $R_n(0)$.*

Proof   For the sake of simplicity, let us suppose that $0 < \alpha < 2$: we may then take $k = 1$ in (ii). From (i), we deduce

$$||(I - T_1)f||_q \leq C\left(||A^{\alpha/2}(I - T_1)f||_2 + ||(I - T_1)f||_2\right)$$
$$\leq C||A^{\alpha/2}f||_2 + ||(I - T_1)f||_2,$$

where $q = 2n/(n - 2\alpha)$. Now (ii) follows from the fact that, if $0 < \alpha < 2$,

$$||(I - T_1)f||_2 \leq C||A^{\alpha/2}f||_2.$$

This may be seen by writing

$$I - T_1 = \int_0^1 AT_t \, dt,$$

and using the analyticity of T_t.

Let us now suppose (ii) and observe that

$$T_t(I - T_1) = A^{-\alpha/2}(I - T_1)A^{\alpha/2}T_t.$$

Since $||A^{\alpha/2}T_t||_{2\to 2} \leq Ct^{-\alpha/2}$, this gives $||T_t(I - T_1)||_{2\to q} \leq Ct^{-\alpha/2}$. It follows that

$$||T_t f||_q \leq C\left(t^{-\alpha/2}||f||_2 + ||T_{t+1}f||_q\right)$$
$$\leq C(t^{-\alpha/2} + 1)||f||_2,$$

since $||T_1||_{2\to q} < +\infty$. Now, applying II.2.2 to $e^{-t}T_t$, we obtain (iii).

To show that (iii) implies (i), use the subordinate semigroup of order $\alpha/2 \in \,]0,1[$ of the semigroup T_t, which we denote by $T_t^{\alpha/2}$. Since T_t satisfies $R_n(0)$, $T_t^{\alpha/2}$ satisfies $R_{2n/\alpha}(0)$ (see Section II.3) and $e^{-t}T_t^{\alpha/2}$ satisfies $R_{2n/\alpha}$. Now one can apply II.2.7 to $e^{-t}T_t^{\alpha/2}$, to obtain that $(I + A^{\alpha/2})^{-1}$ is bounded from L^2 to $L^{2(2n/\alpha)/[(2n/\alpha)-2]}$, hence that

$$||f||_{2n/(n-2\alpha)} \leq C(||A^{\alpha/2}f||_2 + ||f||_2), \quad \forall f \in \mathcal{D}.$$

This ends the proof of II.4.2.

Concerning $R_n(\infty)$, we have

II.4.3 Theorem *Suppose that T_t is equicontinuous on L^1 and L^∞, bounded analytic on L^2, and such that $\|T_1\|_{1\to\infty} < +\infty$, $t > 0$. Let n and α be such that $0 < \alpha < n/2$. Then the following properties are equivalent:*

(i) $\|f\|_{2n/(n-2\alpha)} \le C(\|A^{\alpha/2}f\|_2 + \|A^{\alpha/2}f\|_{2n/(n-2\alpha)})$, *for all $f \in \mathcal{D}$;*

(ii) $\|T_1 f\|_{2n/(n-2\alpha)} \le C\|A^{\alpha/2}f\|_2$, *for all $f \in \mathcal{D}$;*

(iii) T_t *satisfies $R_n(\infty)$.*

Proof Put $q = 2n/(n-2\alpha)$. To deduce (ii) from (i), we simply use

$$\|A^{\alpha/2}T_1 f\|_q \le C\|A^{\alpha/2}f\|_2, \quad \forall f \in \mathcal{D},$$

which comes from the hypothesis $\|T_1\|_{1\to\infty} < +\infty$.

To deduce (iii) from (ii), we write

$$T_{1+t} = T_1 A^{-\alpha/2} A^{\alpha/2} T_t,$$

hence

$$\|T_{1+t}\|_{2\to q} \le \|T_1 A^{-\alpha/2}\|_{2\to q}\|A^{\alpha/2}T_t\|_{2\to 2} = Ct^{-\alpha/2}.$$

The proof of II.2.1 can be immediately adapted, using $\|T_1\|_{1\to\infty} < +\infty$, and one concludes that T_t satisfies $R_n(\infty)$.

Now, Remark II.2.8 applied to $S_t = t^{\alpha/2-1}T_{t+1}$ shows that (iii) implies (ii), and also

$$\|T_1^\beta f\|_q \le C\|A^{\alpha/2}f\|_2, \quad f \in \mathcal{D},$$

where $0 < \beta < 1$ and T_t^β is the subordinate semigroup of order β of T_t. It follows that

$$\|f\|_q \le \|T_1^\beta f\|_q + \|(I - T_1^\beta)f\|_q \le \|A^{\alpha/2}f\|_2 + \|A^{\alpha/2}f\|_q,$$

because, since T_t^β is analytic on L^q,

$$\|(I - T_1^\beta)f\|_q \le \int_0^1 \|A^\beta T_t^\beta f\|_q dt \le C\|A^{\alpha/2}f\|_q.$$

In the last inequality, we have supposed $\alpha/2 < \beta$. The general case can be treated by considering powers of $I - T_1^\beta$.

II.4.4 Remarks (a) More generally, we can show that $R_n(\infty)$ is equivalent to $A^{-\alpha/2}\colon L^p \cap L^q \to L^q$, where $\frac{1}{p} = \frac{1}{q} - \frac{\alpha}{n}$, and $\alpha > 0$. Therefore, by duality and change of indices, it is equivalent to $A^{-\alpha/2}\colon L^p \to L^p + L^q$. We shall use the case $p = 2$, $q = 2n/(n-2)$ in Chapter X.

(b) If one makes stronger assumptions on T_t, for example self-adjoint on L^2 and bounded analytic on L^p, $1 < p < +\infty$, the proofs above become simpler: one doesn't need then to consider subordinate semigroups.

II.5 Symmetric submarkovian semigroups

We briefly recall some classical facts concerning symmetric submarkovian semigroups on L^2.

Let A be an operator of domain $\mathcal{D}(A) \subset L^2$. Then $-A$ is the infinitesimal generator of a symmetric semigroup on L^2 such that $\|e^{-tA}\|_{2\to 2} \le e^{\alpha t}$ if and only if A is self-adjoint, $\mathcal{D}(A)$ is dense in L^2 and $(Af, f) \ge -\alpha \|f\|_2^2$.

Let $Q(f, g)$ be a symmetric bilinear form defined on a subspace $\mathcal{D}$ of L^2. One says that Q is *positive* if $Q(f, f) \ge 0$ and *closed* if for every sequence $(f_n)_{n \in \mathbb{N}} \subset \mathcal{D}$ such that

$$\lim_{n\to\infty} \|f_n - f\|_2 = 0 \quad \text{and} \quad \lim_{n,m\to\infty} Q(f_n - f_m, f_n - f_m) = 0,$$

one has $f \in \mathcal{D}$ and $\lim_{n\to\infty} Q(f_n - f, f_n, f) = 0$. One says that Q is *closable* if it admits a closed extension.

If A is a symmetric operator on a dense subspace $\mathcal{D}$ of L^2, one may associate with it the symmetric bilinear form $Q_A(f, g) = (Af, g)$. If in addition Q_A is positive, it is closable and its minimal closure $\overline{Q}_A$ is associated to a self-adjoint operator $\overline{A}$ which is an extension of A (in fact, the smallest self-adjoint extension of A) called the *Friedrichs extension* of A. From now on we shall not distinguish between A and $\overline{A}$.

Recall that a symmetric semigroup T_t on L^2 is said to be *submarkovian* if

$$f \in L^2, 0 \le f \le 1 \Rightarrow 0 \le T_t f \le 1.$$

Such a semigroup acts on the L^p spaces, $1 \le p \le \infty$, and satisfies $\|T_t\|_{p\to p} \le 1$.

Symmetric submarkovian semigroups on L^2 may be characterized through properties of the associated symmetric bilinear form. A positive symmetric bilinear form Q defined on $\mathcal{D} \subset L^2$ is said to be a *Dirichlet form* if for all $g \in \mathcal{D}$, for all $f \in L^2$ such that $|f| \le |g|$ and $|f(x) - f(y)| \le |g(x) - g(y)|$, we have $f \in \mathcal{D}$ and $Q(f, f) \le Q(g, g)$.

If $T_t = e^{-tA}$ is a symmetric submarkovian semigroup on L^2 then the associated bilinear form $Q(f, g) = (A^{\frac{1}{2}}f, A^{\frac{1}{2}}g)$, $f, g \in \mathcal{D}(A^{\frac{1}{2}})$ is a closed Dirichlet form with dense domain in L^2. Conversely, given a closed Dirichlet form Q with dense domain in L^2, there exists a unique symmetric submarkovian semigroup on L^2 such that $T_t = e^{-tA}$ and $Q(f, g) = (A^{\frac{1}{2}}f, A^{\frac{1}{2}}g)$, $f, g \in \mathcal{D}(A^{\frac{1}{2}})$.

II.5.1 Example Let V be a manifold endowed with a measure m and vector fields $X_1, ..., X_k$, which are formally skew-adjoint on $L^2(V, m)$, i.e. such that

$$\int (X_i f) g \, dm = -\int f(X_i g) \, dm, \quad \forall f, g \in C_0^\infty(V).$$

Then, we can associate with $\Delta = -\sum_{i=1}^k X_i^2$ its Friedrichs extension, which we still denote by Δ, and thus a symmetric semigroup of contractions on L^2,

$H_t = e^{-t\Delta}$; indeed $(\Delta f, g) = (f, \Delta g)$ and

$$(\Delta f, f) = \sum_{i=1}^{k} \int |X_i f|^2 \, dm \geq 0, \quad \forall f, g \in C_0^\infty(V).$$

This last formula also shows that $(\Delta f, g)$ is a Dirichlet form, hence that H_t is symmetric submarkovian. This situation occurs in particular if $V = G$ is a Lie group endowed with its right invariant Haar measure dx and if the vector fields X_i are left invariant. Indeed, let $\varphi \in C_0^\infty(G)$, and X a left invariant vector field on G. Then

$$\int X\varphi(x) \, dx = \int \lim_{t \to 0} \frac{\varphi(xe^{tX}) - \varphi(x)}{t} \, dx = 0$$

since

$$\int \varphi(xe^{tX}) \, dx = \int \varphi(x) \, dx$$

(for the definition of e^{tX}, see any book on Lie group theory or Section III.3 below). Taking $\varphi = fg$, one concludes that X is formally skew-adjoint.

From now on, $T_t = e^{-tA}$ is a symmetric submarkovian semigroup in L^2. Notice that in this setting, the mapping property

$$G_1 = \Gamma(1/2)A^{-\frac{1}{2}} : L^2 \to L^{2n/(n-2)},$$

which results from Theorem II.2.7, and the property D_n, i.e.

$$\|f\|_{2n/(n-2)} \leq C(Af, f), \quad \forall f \in \mathcal{D},$$

are equivalent, since $\|A^{\frac{1}{2}} f\|_2^2 = (Af, f)$, for all $f \in \mathcal{D}$.

We can now sum up several of the above developments in the following theorem.

II.5.2 Theorem *Let T_t be a symmetric submarkovian semigroup, and let $n > 2$. The following are equivalent:*
(i) $\|f\|_{2n/(n-2)}^2 \leq C(Af, f), \quad \forall f \in \mathcal{D};$
(ii) $\|f\|_2^{2+4/n} \leq C(Af, f)\|f\|_1^{4/n}, \quad \forall f \in \mathcal{D};$
(iii) $\|T_t\|_{1 \to \infty} \leq C' t^{-n/2}, \quad \forall t > 0.$

Proof One can invoke alternatively Theorem II.3.4 or Theorem II.3.1 to see that (i) implies (iii). But (i) implies (ii) by Remark II.3.3, (ii) implies (iii) by Theorem II.3.2, and (iii) implies (i) by Theorem II.2.7. and the remark before II.5.2.

II.5.3 Remarks
(a) The theorem holds more generally for symmetric semigroups that are equicontinuous on L^1 and L^∞.

(b) The equivalence between (ii) and (iii) holds for any $n > 0$.

(c) The implication (i) $\Rightarrow$ (ii) is a simple consequence of Hölder's inequality. By contrast, we do not know how to prove (ii) $\Rightarrow$ (i) without considering the semigroup.

We are now going to study the stability of property R_n under perturbations of symmetric submarkovian semigroups. Towards this end, the main tool is the following

II.5.4 Proposition   *Suppose that* e^{-tA} *satisfies, for some* $n > 2$,

$$\|f\|_{2n/(n-2)}^2 \leq C(Af, f) + \gamma\|f\|_2^2, \quad \forall f \in \mathcal{D}.$$

Then it also satisfies

$$\|f\|_{pn/(n-2)}^p \leq Cp^2[4(p-1)]^{-1}\mathrm{Re}\,(Af, f_p) + \gamma\|f\|_p^p, \quad \forall f \in \mathcal{D}.$$

Notice that if A is a second order differential operator, this property follows from a simple integration by parts. The proof of the general case relies on two lemmas.

II.5.5 Lemma   *Let* $\alpha > 0$ *and* $\beta > 0$ *be such that* $\alpha + \beta = 2$ *and* θ_1, θ_2 *be two complex numbers of modulus one. Then*

$$\mathrm{Re}\,[(\theta_1 x^\alpha - \theta_2 y^\alpha)(\overline{\theta}_1 x^\beta - \overline{\theta}_2 y^\beta)] \geq \alpha\beta(x - y)^2, \quad \forall x, y \geq 0.$$

Proof   One has

$$x^\alpha - y^\alpha = \alpha \int_y^x t^{\alpha-1}\, dt,$$

hence

$$(x^\alpha - y^\alpha)(x^\beta - y^\beta) = \alpha\beta \left(\int_y^x t^{\alpha-1}\, dt\right)\left(\int_y^x t^{\beta-1}\, dt\right)$$

$$\geq \alpha\beta \left(\int_x^y t^{(\alpha+\beta)/2-1}\, dt\right)^2 = \alpha\beta(x - y)^2.$$

Besides, it is easy to see that

$$\mathrm{Re}\,[(\theta_1 x^\alpha - \theta_2 y^\alpha)(\overline{\theta}_1 x^\beta - \overline{\theta}_2 y^\beta)] \geq (x^\alpha - y^\alpha)(x^\beta - y^\beta).$$

II.5.6 Lemma   *Let* P *be a symmetric submarkovian operator on* L^2, $\alpha > 0$, $\beta > 0$ *such that* $\alpha + \beta = 2$, ϕ *an* L^2 *positive function such that* ϕ^α *and* ϕ^β *belong to* L^2 *and* θ *a measurable function from* X *to the complex numbers of modulus one. Then*

$$\mathrm{Re}\,((I - P)\theta\phi^\alpha, \theta\phi^\beta) \geq \alpha\beta((I - P)\phi, \phi).$$

Proof It is well known (see References and Comments) that the properties of P imply the existence of a positive measure σ_P on $X \times X$, whose projections on the factor spaces are majorized by the measure ξ and such that

$$(Pf, g) = \int_X \int_X f(x)g(y)\, d\sigma_P(x, y), \quad \forall f, g \in L^2.$$

This gives

$$((I - P)\phi, \phi) = ||\phi||_2^2 - \int_X \int_X \phi^2(x)\, d\sigma_P(x, y)$$
$$+ \frac{1}{2} \int_X \int_X (\phi(x) - \phi(y))^2\, d\sigma_P(x, y)$$

and

$$\mathrm{Re}\,((I - P)\theta\phi^\alpha, \theta\phi^\beta)$$
$$= ||\phi||_2^2 - \int_X \int_X \phi^2(x)\, d\sigma_P(x, y)$$
$$+ \frac{1}{2} \int_X \int_X \mathrm{Re}\,[(\theta(x)\phi^\alpha(x) - \theta(y)\phi^\alpha(y))(\overline{\theta}(x)\phi^\beta(x) - \overline{\theta}(y)\phi^\beta(y))]\, d\sigma_P(x, y).$$

The announced inequality now clearly follows from II.5.5, $\alpha\beta \leq 1$ and from the fact that

$$\int_X \int_X \phi^2(x)\, d\sigma_P(x, y) \leq ||\phi||_2^2, \quad \phi \in L^2.$$

Proof of Proposition II.5.4   Let us change f to $|f|^{p/2}$ in the hypothesis. We obtain

$$||f||_{pn/(n-2)}^p \leq C(A|f|^{p/2}, |f|^{p/2}) + \gamma||f||_p^p.$$

Now, since

$$(Af, g) = \lim_{t \to 0+} \left(t^{-1}(I - T_t)f, g \right), \quad f, g \in \mathcal{D},$$

Lemma II.5.6 holds with A instead of $I - P$. It gives in particular

$$4(p-1)p^{-2}(A|f|^{p/2}, |f|^{p/2}) \leq \mathrm{Re}\,(Af, f_p).$$

This yields the claimed inequality.

As before, let $T_t = e^{-tA}$ be a symmetric submarkovian semigroup and let $S_t = e^{-tB}$ be a semigroup on L^p, $1 \leq p \leq +\infty$. Suppose that there exists a space $\mathcal{D}$ that is dense in L^p, in $\mathcal{D}(A_p)$ and in $\mathcal{D}(B_p)$ for the graph norms, for every p, $1 \leq p < +\infty$. We then have

II.5.7 Theorem   *Suppose that:*
(i) *there exist $n > 2$, $C_1 > 0$ such that*

$$||f||_{2n/(n-2)}^2 \leq C_1(Af, f), \quad \forall f \in \mathcal{D};$$

(ii) *there exist $C_2 > 0$, $\alpha > 0$, γ_p a function of p majorized by a polynomial, such that $\forall p \geq 2$,*

$$\text{Re}\,(Af, f_p) \leq C_2[\text{Re}\,(Bf, f_p) + \alpha\gamma_p\|f\|_p^p], \quad \forall f \in \mathcal{D}.$$

Then there exists $C > 0$ depending only on C_1, C_2, γ_p and n such that:

$$\|S_t f\|_\infty \leq C t^{-n/4}(1 + \alpha t)^{n/4} e^{\alpha t}\|f\|_2, \quad \forall t > 0, \forall f \in L^2.$$

Proof Proposition II.5.4 shows that T_t satisfies

$$\|f\|_{pn/(n-2)}^p \leq C_1 p^2 [4(p-1)]^{-1}\text{Re}\,(Af, f_p), \quad \forall p \geq 2, \forall f \in \mathcal{D}.$$

Together with hypothesis (iii), this gives

$$\|f\|_{pn/(n-2)}^p \leq C_1 C_2 p^2 [4(p-1)]^{-1}\text{Re}\,(Bf, f_p) + \alpha\gamma_p\|f\|_p^p, \quad \forall p \geq 2, \forall f \in \mathcal{D}.$$

We are now in a position to apply Theorem II.3.5, which ends the proof.

II.5.8 Corollary *Suppose that A and B fulfill the hypothesis of II.5.7, and that (ii) also holds if one replaces B by its adjoint B^*. Then*

$$\|S_t f\|_\infty \leq C t^{-n/2}(1 + \alpha t)^{n/2} e^{\alpha t}\|f\|_1, \quad \forall t > 0.$$

Proof We not only have

$$\|S_t\|_{2\to\infty} \leq C t^{-n/4}(1 + \alpha t)^{n/4} e^{\alpha t}, \quad \forall t > 0,$$

but also the same estimate for the adjoint semigroup S_t^*. This gives

$$\|S_t\|_{1\to 2} \leq C t^{-n/4}(1 + \alpha t)^{n/4} e^{\alpha t}, \quad \forall t > 0.$$

Finally

$$\|S_t\|_{1\to\infty} \leq \|S_{t/2}\|_{1\to 2}\|S_{t/2}\|_{2\to\infty} \leq C t^{-n/2}(1 + \alpha t)^{n/2} e^{\alpha t}, \quad \forall t > 0.$$

II.5.9 Remark Suppose that moreover

$$\text{Re}\,(e^{i\varepsilon} Bf, f) \geq -\alpha(1 + |\varepsilon|)\|f\|_2^2, \quad 0 \leq \varepsilon < \theta.$$

Then $S_t = e^{-tB}$ admits a holomorphic extension to the sector

$$S_\theta = \{\zeta \in \mathbb{C}, \text{Re}\,\zeta > 0, |\arg(\zeta)| < \theta\},$$

and a classical Cauchy integral argument shows that, for $0 < \varepsilon < \theta$,

$$\left\|\left(\frac{\partial}{\partial t}\right)^k S_t\right\|_{2\to 2} \leq C_k(\varepsilon) t^{-k} \exp(\alpha(1 + \varepsilon)t), \quad \forall t > 0.$$

Combining this estimate with the conclusion of Corollary II.5.8, we obtain, for $0 < \varepsilon < \theta$,

$$\left\| \left(\frac{\partial}{\partial t} \right)^k S_t \right\|_{1 \to \infty} \leq C_k(\varepsilon) t^{-n/2-k} (1 + \alpha t)^{n/2} \exp(\alpha(1 + \varepsilon)t), \quad \forall t > 0.$$

References and comments

The main features of this chapter are due to N. Varopoulos ([132], [136], [139], [140]). He first developed this theory in the setting of Markov chains ([127], [128], [129], [130], [131], [135]), using ideas of Moser ([90], [91]). He then extended these ideas to continuous time semigroups. For Nash's argument II.3.2, see [95], [140], [20], [28].

It turned out that each of the two sides of the equivalence $R_n \Leftrightarrow D_n$ (cf. II.1) had at least one ancestor: Yoshikawa ([158]) had essentially proved that $R_n \Rightarrow D_n$ and Bénilan ([14]), that $D_n \Rightarrow R_n$; see also [75]. Moreover, Fukushima ([51]) had considered the connection between D_n type properties and estimates on the resolvant which follows also from Remark II.3.3 (c).

For their part, Davies and Simon have studied in [44] the $L^1 - L^\infty$ regularization properties of the Schrödinger semigroup $e^{t(\Delta+V)}$, by means of logarithmic Sobolev inequalities of Gross [58]. Their theory has been generalized since, and is set forth in [42].

The proofs that we give here for Theorems II.3.1, II.3.4 and II.3.5 are due to T. Coulhon ([32], [24], [25], [26]) and those of the extrapolation lemmas II.2.1 and II.2.2, to T. Coulhon and Y. Raynaud. The original extrapolation procedure on H_α spaces on the unit disc goes back to the classical work of Hardy and Littlewood. The localization at infinity in Section II.4 comes from [24]. These techniques have been applied in the context of the Stokes semigroup [30].

Further developments in the direction of Sobolev type imbedding theorems for semigroups can be found in [32], [33], [34], [38], [78], [28].

The facts we used about semigroups are explained in [39] or [98]; concerning analytic semigroups, see also [119] and [159]. The theory of Dirichlet forms, which is due to Beurling and Deny, and the expression of a symmetric submarkovian operator used in II.5.6 can be found in [45], [50], [42]. For Marcinkiewicz's theorem II.2.5, see [118].

CHAPTER III

SYSTEMS OF VECTOR FIELDS
SATISFYING HÖRMANDER'S CONDITION

In this chapter, we consider a family $X_1, ..., X_k$ of vector fields on a manifold V, and we suppose that, at every point of V, the Lie algebra they generate is the whole tangent space; this is called Hörmander's condition.

Let $\Delta = -\sum_{i=1}^{k} X_i^2$; it will be the basic operator in this book. The aim of the present chapter is to show, or recall, three important consequences of Hörmander's condition: Δ is hypoelliptic (Section 1), a Harnack principle holds for positive solutions of $\frac{\partial}{\partial t} + \Delta$ (Section 2), and a distance on V can naturally be associated with Δ (Section 4). In Section 3, we present a technical variation on the theme of the Campbell-Hausdorff formula. It is used in Section 4 and in Chapter V.

Chapters IV, V, VIII and IX directly depend on the present one. However, uniformity considerations may be left aside on a first reading, since they will be used only in Chapter V.

III.1 Hörmander's condition and hypoellipticity

Let V be a C^∞ connected manifold of dimension N. Let us consider a system $\mathbf{X} = \{X_1, ..., X_k\}$ of C^∞ vector fields on V. We shall denote by $\mathcal{L}_{\mathbf{X}}$ the Lie algebra generated by $\mathbf{X}$ in the Lie algebra of C^∞ vector fields on V.

If $x \in V$ and if $T_x V$ is the tangent space to V at x, $\mathcal{L}_{\mathbf{X}}(x)$ will denote the linear subspace of $T_x V$ obtained by evaluating at x the elements of $\mathcal{L}_{\mathbf{X}}$.

III.1.1 Definition   A system $\mathbf{X}$ of C^∞ vector fields on V is said to satisfy Hörmander's condition, or to be a Hörmander system of vector fields, if for every $x \in X$, $\mathcal{L}_{\mathbf{X}}(x) = T_x(V)$.

Hörmander has shown that, under this condition, the second order differential operators associated with $\mathbf{X}$ are hypoelliptic. Before stating precisely Hörmander's theorem, let us recall that a differential operator D on V is said to be hypoelliptic if $Du \in C^\infty(\Omega)$ implies $u \in C^\infty(\Omega)$ for every distribution u on V and every open subset Ω of V.

III.1.2 Theorem   *If $\mathbf{X} = \{X_1, ..., X_{k+1}\}$ is a Hörmander system of vector fields on V, and if $a \in C^\infty(V)$, the differential operator $D = \sum_{i=1}^{k} X_i^2 + X_{k+1} + a$ is hypoelliptic.*

III.1.3 Corollary   *Given an open subset Ω of V, a compact subset K of V and $p \in \mathbb{N}$, there exists a constant C such that every solution v of $Dv = 0$ on V satisfies*

$$\|v\|_{C^p(K)} \leq C \int_\Omega |v|$$

(the integral is taken with respect to a C^∞ non-vanishing measure on V).

Theorem III.1.2 may be considered as classical, and we do not give a proof here. The corollary follows from the closed graph theorem.

Uniformity matters

Consider now a system of vector fields depending on one parameter: $\mathbf{X}^s = \{X_1^s, ..., X_{k+1}^s\}$, where $s \in S \subset \mathbb{R}$. We want to give conditions on $\mathbf{X}^s$ which yield a version of Corollary III.1.3 that is uniform with respect to s.

We shall say that a set B of C^∞ vector fields on V is bounded if, for every chart ϕ of V, every compact K included in the domain of ϕ and every integer p, there exists C such that

$$\forall X \in B, \forall i \in \{1, ..., N\}, \|a_i\|_{C^p(K)} \leq C,$$

where $a_i \in C^\infty(V)$ is the ith coordinate of X in the basis $\frac{\partial}{\partial x_1}, ..., \frac{\partial}{\partial x_N}$ induced by the chart ϕ. The condition that $\{X_i^s \mid i \in \{1, ..., k+1\}, s \in S\}$ should be a bounded set of fields does not suffice to ensure the hoped-for uniformity: one also has to prevent the fields from being in a certain sense too small. Let

$$\mathcal{I}(k) = \bigsqcup_{\alpha \in \mathbb{N}^*} \{1, ..., k\}^\alpha$$

be the set of multi-indices I with values in $\{1, ..., k\}$, of arbitrary length $|I| = \alpha$. For $I = \{i_1, ..., i_\alpha\} \in \mathcal{I}(k)$, let

$$\mathbf{X}_I = [X_{i_1}, [X_{i_2}, ..., [X_{i_{\alpha-1}}, X_{i_\alpha}]...]].$$

In this notation, a system $\mathbf{X} = \{X_1, ..., X_k\}$ is a Hörmander system if, for every point x of V, there exists a subset J_x of $\mathcal{I}(k)$ such that $\{\mathbf{X}_I(x) \mid I \in J_x\}$ is a basis of T_xV.

III.1.4 Definition   The system of vector fields $\mathbf{X}^s = \{X_1^s, ..., X_{k+1}^s\}$ satisfies the Hörmander condition uniformly with respect to $s \in S \subset \mathbb{R}$ if:

(1) $\{X_i^s \mid s \in S, i \in \{1, ..., k+1\}\}$ is a bounded set of vector fields.

(2) For all x in V, there exists $J_x \in \mathcal{I}(k+1)$ such that:

(i) $\{X_I^s \mid I \in J_x\}$ is a basis of T_xV, for all $s \in S$.

(ii) In a chart U containing x, the transition matrices from $\{X_I^s \mid I \in J_x\}$ to $\{\frac{\partial}{\partial x_1}, ..., \frac{\partial}{\partial x_N}\}$ and their inverses have uniformly bounded coefficients in $C^\infty(U)$ with respect to s.

III.1.5 Theorem   *Assume that* $\mathbf{X}^s$ *satisfies the Hörmander condition uniformly with respect to* $s \in S$, *that* $\{a_s \mid s \in S\}$ *is a bounded subset of* $C^\infty(V)$, *and set* $D^s = \sum_{i=1}^k (X_i^s)^2 + X_{k+1}^s + a_s$. *Then, for every open set* $\Omega \subset V$, *for every compact* $K \subset \Omega$ *and for every integer* p, *there exists a constant* C *such that*

$$\|v\|_{C^p(K)} \leq C \int_\Omega |v|,$$

for every distribution v on V such that there exists $s \in S$ for which $D^s v = 0$.

Unfortunately, to our knowledge, the only way to obtain Theorem III.1.5 is to follow the proof of III.1.2 (see References and Comments) step by step, keeping the constants under control.

III.2 Harnack inequalities

The prototype of a Harnack inequality is the following result, which holds in the Euclidean space: there exists a constant C such that, for every $R > 0$, every $x \in \mathbb{R}^n$ and every positive harmonic function u in the ball centred at x of radius R, we have

$$\sup\{u(y) \mid y \in B(x, R/2)\} \leq C \inf\{u(y) \mid y \in B(x, R/2)\}.$$

This elliptic result admits a parabolic version: there exists a constant C such that, for every $T > 0$, for every $x \in \mathbb{R}^n$ and for every positive solution of $\left(\frac{\partial}{\partial t} + \Delta\right) u = 0$ in $]0, T[\times B(x, \sqrt{T})$, one has

$$\sup\{u(T/3, y) \mid y \in B(x, \sqrt{T}/2)\} \leq C \inf\{u(T/2, y) \mid y \in B(x, \sqrt{T}/2)\}.$$

In this inequality, we can replace u in the left hand side by

$$|\frac{\partial^n}{\partial t^n} \frac{\partial^\alpha}{\partial x^\alpha} u|,$$

provided we insert a factor of $T^{-n-|\alpha|/2}$ in the right hand side. In addition, we clearly see the necessity of comparing $\sup u(t_0, .)$ with $\inf u(t_1, .)$ for $t_0 < t_1$ (some time has to elapse), by looking at the family of solutions

$$t^{-n/2} \exp(-\frac{|x - a|^2}{4t}),$$

where a runs through $\mathbb{R}^n$.

This parabolic Harnack inequality in $\mathbb{R}^n$ contains indeed two levels of information: on the one hand, the fact that, if x and y vary in a compact set in $\mathbb{R}^n$ and for $T > 0$ fixed, the quantity $u(T/3, x)/u(T/2, y)$ is bounded above, independently of the positive solution u of $\frac{\partial}{\partial t} + \Delta$; and on the other hand, the fact that this estimate above is uniform in T, provided x and y stay in a ball of radius $\sqrt{T}$. One passes very easily from the first level to the second, thanks to the dilation structure of $\mathbb{R}_+ \times \mathbb{R}^n$ and the homogeneity of the operator $\frac{\partial}{\partial t} + \Delta$.

Things are different if one wants to obtain analogous inequalities for positive solutions of

$$\frac{\partial}{\partial t} - \sum_{i=1}^k X_i^2$$

on $\mathbb{R}_+ \times V$, where V is a manifold, under the hypothesis that $\{X_1, ..., X_k\}$ is a Hörmander system. We shall see in this section that the "T-fixed" Harnack inequality follows from the hypoellipticity of the considered operator and from the construction of an appropriate Green function. As for the "scaling in T" of this inequality, it can only be performed, in this setting, for T small; it is difficult and will be achieved only in Chapter V. The developments about uniformity that we began in Section III.1 and that we shall pursue hereafter are means to this end. For large T, the scaling will be achieved, in the setting of polynomial growth Lie groups, in Chapter VIII. Note that, in contrast, the scaling of the Harnack inequality will be performed easily in the particular case of nilpotent Lie groups, in Chapter IV.

Thus, our present purpose is the following statement:

III.2.1 Theorem   *Let* $\mathbf{X} = \{X_1, ..., X_k\}$ *be a Hörmander system of vector fields on the manifold* V, Y *a* C^∞ *vector field on* V, *and a a* C^∞ *function on* V. *Let* D *be the differential operator*

$$\sum_{i=1}^{k} X_i^2 + Y - a - \frac{\partial}{\partial t},$$

defined on $\mathbb{R} \times V$. *Let* $I =]\alpha, \beta[\subset \mathbb{R}$, Ω *an open relatively compact subset of* V, K *a compact subset of* Ω, t_1, t_2 *such that* $\alpha < t_1 < t_2 < \beta$, $J \in \mathcal{I}(N)$, $m \in \mathbb{N}$.

Then there exists a constant C *such that every positive solution* u *of* $Du = 0$ *in* $I \times \Omega$ *satisfies*

$$\sup_{x \in K} \left| \left(\frac{\partial}{\partial t} \right)^m \left(\frac{\partial}{\partial x} \right)^J u(t_1, x) \right| \leq C \inf_{x \in K} u(t_2, x).$$

(Spatial derivatives are taken with respect to some fixed system of local coordinates.)

We are now going to sketch the main steps of the proof; we shall state some intermediate results, in order to be able to come back to the uniformity problem.

Notice first that the system $\{X_1, ..., X_k, Y - \frac{\partial}{\partial t}\}$ is Hörmander on $\mathbb{R} \times V$. We will thus be in position to apply III.1.3. Consider the operator $\widetilde{D} = D - \lambda$, for λ such that $a + \lambda \geq 1$; this shift ensures the existence of an integrable Green function. The Dirichlet problem associated with $\widetilde{D}$ on an open set $W \subset I \times V$ consists of searching for the solutions u of

$$\widetilde{D}(u) = -f \text{ in } W$$
$$u = \phi \text{ in } \partial W$$

where the data f and ϕ are continuous on $\overline{W}$ and ∂W respectively.

If W is an open set such that for every pair (f, ϕ) as above, the Dirichlet problem has a unique solution (one then says that W is regular for the Dirichlet problem associated with $\widetilde{D}$). We may then consider the operator H, which associates with a continuous function ϕ on ∂W, the solution $u = H\phi$ of the Dirichlet problem with data $(0, \phi)$, and the operator G which associates with a continuous function f on $\overline{W}$, the solution $u = Gf$ of the Dirichlet problem with data $(f, 0)$. We call H the Poisson operator and G the Green operator. The solution of the Dirichlet problem with data (f, ϕ) may of course be written $Gf + H\phi$. Any C^2 function u on $\overline{W}$ may be written $u = -G\widetilde{D}u + H(u|_{\partial W})$. Now, since $a + \lambda \geq 0$, the maximum principle holds in our setting, and H is a positive operator; thus, if u is a positive solution of $Du = 0$, we have

$$u \geq -G\widetilde{D}u \geq Gu.$$

An essential tool in the proof of III.2.1 is a result on propagation of minima.

III.2.2 Lemma   *Let W be an open subset of $I \times V$ and u a positive solution of $\widetilde{D}u = 0$ in W. Let Z be a C^∞ vector field on V and $\gamma(s)$ a path in W, satisfying $\gamma'(s) = Z(\gamma(s)) - \frac{\partial}{\partial t}$. If $u(\gamma(s_0)) = 0$, then $u(\gamma(s)) = 0$, for all $s > s_0$.*

It follows immediately from this lemma that a positive solution

$$(t, x) \mapsto u(t, x) \quad \text{of} \quad \widetilde{D}u = 0$$

in an open subset W of $I \times V$, which vanishes in $(t_0, x_0) \in W$, vanishes everywhere in $(]-\infty, t_0[\times V) \cap W$.

We are now going to work locally, in an open subset U of $I \times V$ that we shall identify through a chart with an open subset of $\mathbb{R}^{N+1}$. Hence, $\widetilde{D}$ can be written

$$\widetilde{D} = \sum_{i,j}^{N} a_{ij}(x)\frac{\partial^2}{\partial x_i \partial x_j} + a_i(x)\frac{\partial}{\partial x_i} - a(x) - \lambda - \frac{\partial}{\partial t}.$$

If F is a closed subset of U, a vector v of $\mathbb{R}^{N+1}$ is said to be normal to F at $\gamma_0 \in F$ if there exists an open ball B contained in $U\backslash F$ such that $\gamma_0 \in \partial B$ and $v = \alpha(\gamma_1 - \gamma_0)$, where γ_1 is the centre of B and $\alpha > 0$.

We shall say that an open set W of U has the property (B) if, at every point $\gamma = (t, x)$ of ∂W, there exists a vector $v = (u, v_1, ..., v_N)$ which is normal to $\overline{W}$ at γ and such that

$$\sum_{i,j=1}^{N} a_{ij}(x)v_i v_j > 0.$$

It can be shown that V admits a basis of open sets satisfying (B). Moreover

III.2.3 Proposition   *Let W be an open set satisfying (B). Then W is regular for the Dirichlet problem associated with $\widetilde{D}$ and there exists a function*

$g: W \times W \to \mathbb{R}^+$, C^∞ *outside the diagonal, such that, for every continuous function f on $\overline{W}$,*

$$Gf(\gamma) = \int_W g(\gamma, \xi) f(\xi)\, d\xi, \quad \forall \gamma \in W;$$

g *is called the Green function associated with $\widetilde{D}$ on W.*

If $\widetilde{D}^* = \sum_{i=1}^k X_i^2 - Y - a - \lambda + \frac{\partial}{\partial t}$, *the Green function associated with $\widetilde{D}^*$ is given by $g^*(\gamma, \xi) = g(\xi, \gamma)$. Moreover, we have*

$$\int_W g(\gamma, \xi)\, d\xi \leq 1.$$

Let us now come back to the proof of III.2.1. We may suppose that $[\alpha, \beta] \times \overline{\Omega} \subset W$, where W is an open subset of $I \times V$ satisfying (B). Indeed, Harnack's inequality is a local property.

Let $\varepsilon > 0$ be such that $]t_1 - \varepsilon, t_1 + \varepsilon[\times \Omega \subset W$, $m \in \mathbb{N}$ and $J \in \mathcal{I}(N)$; Corollary III.1.3 yields a constant C such that every positive solution u of $Du = 0$ in $I \times \Omega$ satisfies

$$\sup_{x \in K} \left| \left(\frac{\partial}{\partial t}\right)^m \left(\frac{\partial}{\partial x}\right)^J u(t_1, x)\right| \leq C \int_{]t_1 - \varepsilon, t_1 + \varepsilon[\times \Omega} u(\xi)\, d\xi. \tag{1}$$

Moreover, since g^* is the Green function associated with $\widetilde{D}$, we have

$$\widetilde{D}_\gamma g^*(\xi, \gamma) = -\delta_\xi,$$

that is

$$\widetilde{D}_\gamma g(\gamma, \xi) = -\delta_\xi, \quad \forall \xi \in W.$$

Thus $v = g(., \xi)$ is a positive solution of $\widetilde{D}v = 0$ on $W/\{\xi\}$.

Choose ε such that $t_1 + \varepsilon < t_2$, and fix ξ in $[t_1 - \varepsilon, t_1 + \varepsilon] \times \overline{\Omega}$; $g(., \xi)$ cannot vanish at $\gamma_0 \in \{t_2\} \times K$, because otherwise, according to Lemma III.2.2, it would vanish on $(]-\infty, t_2[\times V) \cap W$, an open subset containing ξ, and could not satisfy $\widetilde{D}_\gamma g(\gamma, \xi) = -\delta_\xi$. Finally

$$\inf\{g(\gamma, \xi) \mid \gamma \in \{t_2\} \times K, \xi \in [t_1 - \varepsilon, t_1 + \varepsilon] \times \overline{\Omega}\} = C^{-1} > 0$$

and

$$\sup_{x \in K} \left| \left(\frac{\partial}{\partial t}\right)^m \left(\frac{\partial}{\partial x}\right)^J u(t_1, x)\right| \leq C^2 \int_{]t_1 - \varepsilon, t_1 + \varepsilon[\times \Omega} g(\gamma, \xi) u(\xi)\, d\xi.$$

$$\leq C^2 \int_W g(\gamma, \xi) u(\xi)\, d\xi = C^2 (Gu)(\gamma)$$

$$\leq C^2 u(\gamma),$$

and this holds for all $\gamma \in \{t_1\} \times K$. Q. E. D.

Uniformity matters

Resuming the approach of the previous section, we are going to show:

III.2.4 Theorem   *Let $\mathbf{X}^s = \{X_1^s, ..., X_k^s\}$ be a family of systems of C^∞ vector fields on V satisfying the Hörmander condition uniformly with respect to $s \in S \subset \mathbb{R}$, Y^s a family of fields which is bounded for the C^∞ topology, and a_s a bounded family of C^∞ functions on V. Set*

$$D^s = \sum_{i=1}^{k}(X_i^s)^2 + Y^s - a_s - \frac{\partial}{\partial t}.$$

Let $I =]\alpha, \beta[\subset \mathbb{R}$, K a compact subset of V, t_1, t_2 such that $\alpha < t_0 < t_1 < \beta$, $m \in \mathbb{N}$ and $J \in \mathcal{I}(N)$. Then there exists a constant C such that, for every $s \in S$ and every positive solution u of $D^s u = 0$ in $I \times V$, we have

$$\sup_{x \in K} |\left(\frac{\partial}{\partial t}\right)^m \left(\frac{\partial}{\partial x}\right)^J u(t_1, x)| \le C \inf_{x \in K} u(t_2, x).$$

Proof We can suppose that $S = \mathbb{N}^*$, that $\lim_{s \to +\infty} X_i^s = X_i^\infty$, $i = 1, ..., k$, $\lim_{s \to +\infty} Y^s = Y^\infty$, and $\lim_{s \to +\infty} a_s = a_\infty$ exist for the C^∞ topology. It is then clear that, for large enough s, V admits a basis of open sets satisfying the conditions (B^s) relative to D^s. We can thus reduce again the proof to the case where $[t_0, t_1] \subset]\alpha, \beta[$, $K \subset \Omega$ open, with $[\alpha, \beta] \times \overline{\Omega} \subset W$, where W is an open subset of $I \times V$ satisfying (B^s), for all $s \in \mathbb{N}^*$.

Now, it is easy to see that the system of fields $\{X_1^s, ..., X_k^s, Y^s - \frac{\partial}{\partial t}\}$ is Hörmander uniformly on $\mathbb{R} \times V$. Thus, according to Theorem III.1.5, given $m \in \mathbb{N}$ and $J \in \mathcal{I}(N)$, there exists C such that, for every $s \in \mathbb{N}^*$ and every positive solution u of $D^s u = 0$ in $I \times V$, we have

$$\sup_{x \in K} |\left(\frac{\partial}{\partial t}\right)^m \left(\frac{\partial}{\partial x}\right)^J u(t_1, x)| \le C \int_{]t_1 - \varepsilon, t_1 + \varepsilon[\times \Omega} u(\xi)\, d\xi$$

where ε is such that $[t_1 - \varepsilon, t_1 + \varepsilon] \times \overline{\Omega} \subset W$ and $t_1 + \varepsilon < t_2$. In order to finish as in III.2.1, it suffices to show that

$$\inf\{g^s(\gamma, \xi) \mid \gamma \in \{t_2\} \times K, \xi \in [t_1 - \varepsilon, t_1 + \varepsilon] \times \overline{\Omega}\} \ge C^{-1} > 0,$$

with C independent of s, where g^s is the Green function associated with $\widetilde{D}^s = D^s - \lambda$; here λ is chosen so that $a_s + \lambda \ge 1$, $\forall s$.

If this is not the case, one can always suppose that there exists a family $(\gamma_s, \xi_s)_{s \in \mathbb{N}^*}$ such that

$$g^s(\gamma_s, \xi_s) \le \frac{1}{s}, \quad \gamma_s \in \{t_2\} \times K, \quad \xi_s \in [t_1 - \varepsilon, t_1 + \varepsilon] \times \overline{\Omega}.$$

We can also suppose that $\gamma_s \to \gamma_\infty$ and $\xi_s \to \xi_\infty$. Finally, since for all s

$$\int_W g^s(\gamma, \xi)\, d\gamma \le 1,$$

we can assume that $u_s = g^s(., \xi_s)$ converges to u_∞ in the distributions sense.

Letting s tend to infinity in the equality $D_\gamma^s u_s = -\delta_{\xi_s}$ gives, with obvious notation, $D_\gamma^\infty u_\infty = -\delta_{\xi_\infty}$, hence $u_\infty = g^\infty(., \xi_\infty)$. Applying again III.1.5 and the boundedness in $L^1(W)$ of the family u_s, one sees that u_s and u_∞ are in fact C^∞ on $\{t_2\} \times K$, and that one can assume that u_s tends uniformly to u_∞ on $\{t_2\} \times K$. Finally, $u_\infty(\gamma_\infty) = \lim u_s(\gamma_s) = \lim g^s(\gamma_s, \xi_s) = 0$, which is inconsistent with the fact that u_∞ cannot vanish at a point of $\{t_2\} \times K$. This ends the proof of Theorem III.2.4.

Let us now sketch an alternative proof of Theorems III.2.1 and III.2.4. For simplicity, assume that $a \geq 0$. Consider the minimal heat kernel h_t^Ω associated with D on Ω, with Dirichlet boundary conditions. There are several ways to construct it: probability theory, potential theory or semigroup theory (see References and Comments). Let u be a positive solution of $Du = 0$ on $I \times \Omega$. By minimality of h_t^Ω, we have

$$u(t_2, x) \geq \int_V h_{t_2-t}^\Omega(x, y) u(t, y)\, dy, \quad \forall t \in]0, t_2[.$$

It follows that

$$u(t_2, x) \geq \int_K h_{t_2-t}^\Omega(x, y) u(t, y)\, dy$$
$$\geq \frac{1}{s_2 - s_1} \int_{K \times [s_1, s_2]} h_{t_2-t}^\Omega(x, y) u(t, y)\, dt\, dy,$$

for $0 < s_1 < s_2 < t_1$. Hence, for $\zeta \in K$,

$$u(t_2, \zeta) \geq C \int_{K \times [s_1, s_2]} u(t, y)\, dt\, dy,$$

where

$$C = \frac{1}{s_2 - s_1} \inf\{h_{t_2-t}^\Omega(x, y) | x \in K, y \in K, t \in [s_1, s_2]\}$$

is positive and independent of u.

Now, if $t_1 \in]s_1, s_2[$,

$$\sup_{x \in K} \left| \left(\frac{\partial}{\partial t}\right)^m \left(\frac{\partial}{\partial x}\right)^J u(t_1, x) \right|$$

is obviously dominated by a C^p norm of u on $K \times [s_1, s_2]$, which is in turn, thanks to Lemma III.1.3, dominated by

$$\int_{K \times [s_1, s_2]} u(t, y)\, dt\, dy.$$

Finally, for all $\zeta \in K$

$$\sup_{x \in K} \left| \left(\frac{\partial}{\partial t}\right)^m \left(\frac{\partial}{\partial x}\right)^J u(t_0, x) \right| \leq C u(t_2, \zeta),$$

which ends the proof.

Now for III.2.4, i.e. the case where $\mathbf{X}^s$ depends on a parameter $s \in S$. Again, suppose for simplicity that $a_s \geq 0$. Let $h_t^{\Omega,s}$ be the heat kernel with Dirichlet boundary conditions on Ω, associated to D^s. The same approach as in III.2.1 will give the theorem, provided that

$$\exists C > 0 \text{ such that } h_t^{\Omega,s}(x,y) \geq C, \quad \forall s \in S, \forall t \in [t_3,t_4], \forall (x,y) \in K \times K.$$

Suppose this is not the case: there exist sequences $s_n \to s_\infty \in \overline{S}$, $t_n \to t_\infty \in [t_3,t_4]$, $(x_n,y_n) \to (x_\infty,y_\infty) \in K \times K$, such that $h_{t_n}^{\Omega,s_n}(x_n,y_n) \to 0$.

By extracting subsequences, one can assume that $X_i^{s_n} \to X_i^\infty$, $i = 1, ..., k$, $Y^{s_n} \to Y^\infty$, and $a_{s_n} \to a_\infty$, for the C^∞ topology. Set $D^\infty = \sum_{i=1}^k (X_i^\infty)^2 + Y^\infty - a_\infty - \frac{\partial}{\partial t}$.

Finally, $D^{s_n} h^{\Omega,s_n}(.,y_n) = 0$ for all n, on $I \times \Omega$, and

$$\int_{I \times \Omega} h_t^{\Omega,s_n}(x,y_n)\, dx \leq |I|, \text{ for all } n,$$

therefore Theorem III.1.5 tells us that the family of functions $h^{\Omega,s_n}(.,y_n)$ is bounded for the C^∞ topology on $I \times \Omega$. Thus we can suppose that it converges in the C^∞ sense to a function u on $I \times \Omega$ that obviously satisfies $D^\infty u = 0$ and $u(t_\infty,x_\infty) = 0$. By Bony's maximum principle, this implies $u(t_1,x) = 0$, for $t_1 < t_\infty$, for all $x \in \Omega$; in other words, $h_{t_1}^{\Omega,s_n}(x,y_n) \to 0$ uniformly in x.

But this is impossible; indeed let $\phi \not\equiv 0$ be a fixed, positive and compactly supported function in Ω, and let H_t^s be the heat semigroup of kernel $h_t^{\Omega,s}$. We have

$$|H_{t_1}^s \phi(x_\infty) - \phi(x_\infty)| \leq \int_0^{t_1} |H_t^s D^s \phi(x_\infty)|\, dt$$
$$\leq t_1 \|D^s \phi\|_\infty,$$

and this last quantity only depends on the L^∞ norms of the coefficients of D^∞, which are supposed to be bounded. Therefore, for small enough t_1, $H_{t_1}^s \phi(x_\infty)$ is bounded from below in s. This prevents $h_{t_1}^{\Omega,s_n}(.,y_n)$ from tending to zero uniformly, which is the desired contradiction.

III.3 The exponential map

In Chapter V, we shall need, in the setting of manifolds, tools from the differential calculus analogous to those given in the setting of Lie groups by the classical exponential map. We are going to develop such tools in this section.

Let V be a C^∞ manifold, K a compact subset of V and B a set of C^∞ vector fields bounded for the C^∞ topology. From Cauchy's theorem for differential equations, there exists $\delta > 0$ such that, for every $x \in K$, and every $X \in B$, the equation $df(\frac{d}{dt}) = X$, $f(0) = x$, has a unique solution defined for $|t| > \delta$; we shall denote this solution by $E_X(t,x)$. Thanks to uniqueness, it is clear that $E_X(t, E_X(s,x)) = E_X(t+s,x)$ and $E_{\lambda X}(t,x) =$

$E_X(\lambda t, x)$, for small enough s, t and λ ; $E_X(t, x)$ depends on X in a C^∞ way.

We define the map $\exp(X)$ by $\exp(X)(x) = E_X(1, x)$, when the right hand side exists. It follows from the above remarks that, for given X and x, $\exp(tX)$ is always defined in the neighbourhood of x for t small enough, and that it is a diffeomorphism, whose inverse is $\exp(-tX)$. If X and Y are two C^∞ vector fields, $\exp(sY)\exp(tX)$ is locally defined as a composed mapping, for small enough s and t .

If f is a C^∞ function defined around a point $x \in V$, and if X is a C^∞ vector field on V, the function $F(t) = f(\exp(tX)(x))$ is defined and C^∞ around $t = 0$. The computation of the successive derivatives of F through the differential equation that defines the exponential map shows that the Taylor series of F at the origin is $\sum_{p=0}^\infty \frac{t^p}{p!}(X^p f)(x)$, which we write $(e^{tX}f)(x)$.

Let K be a compact subset of V and $\mathbf{Y} = (Y_1, ..., Y_n)$ a system of C^∞ vector fields on V. For $u = (u_1, ..., u_n) \in \mathbb{R}^n$, set $u \cdot \mathbf{Y} = \sum_{i=1}^n u_i Y_i$. It follows from the above that there exists $\delta > 0$ such that, for every $x \in K$, the mapping $\exp_{\mathbf{Y},x} : u \to \exp(u \cdot \mathbf{Y})(x)$, is defined on

$$U = \{(u_1, ..., u_n) \mid |u_i| < \delta, i = 1, ..., n\}.$$

Moreover, the mapping $\exp_{\mathbf{Y},x}$ is C^∞ on U and if f is C^∞ in the neighbourhood of x, the Taylor series of $f(\exp_{\mathbf{Y},x}(v))$ is given by

$$\sum_{p=0}^{+\infty} \frac{1}{p!}[(u \cdot \mathbf{Y})^p f](x),$$

which we write $(e^{u \cdot \mathbf{Y}}f)(x)$. One has in particular

$$d\exp_{\mathbf{Y},x}(0)\left(\frac{\partial}{\partial u_i}\right) = Y_i(x), \quad i = 1, ..., n.$$

If moreover $n = N$ and $\{Y_1(x), ..., Y_N(x)\}$ is a basis of $T_x V$, the mapping $\exp_{\mathbf{Y},x}$ is a diffeomorphism from a neighbourhood of 0 in $\mathbb{R}^N$ to a neighbourhood of x in V. We shall call exponential local coordinates built on $\mathbf{Y}$ at x the local coordinates around x given by $\exp_{\mathbf{Y},x}^{-1}$.

Again, let $\mathbf{Y} = \{Y_1, ..., Y_n\}$ be a system of C^∞ vector fields on V, let $x \in V$, and let $\exp_{\mathbf{Y},x}$ be the associated exponential mapping. We are now going to give an asymptotic expansion, for u in a neighbourhood of 0, of $d\exp_{\mathbf{Y},x}(v)(\frac{\partial}{\partial u_1})$; $(\mathrm{ad}\, Y)X$ will denote $[Y, X]$, $|u|$ the Euclidean length of $u \in \mathbb{R}^n$, and $\|E(u)\|$ the length of a vector $E(u)$ belonging to $T_{\exp_{\mathbf{Y},x}(u)}V$, induced by some fixed system of local coordinates around x.

III.3.1 Proposition *For $i = 1, ..., n$, we have*

$$d\exp_{\mathbf{Y},x}(u)\left(\frac{\partial}{\partial u_i}\right) \sim \sum_{m=0}^{+\infty} \frac{(-1)^m}{(m+1)!}[(\mathrm{ad}\,(u \cdot \mathbf{Y}))^m Y_i](\exp_{\mathbf{Y},x}(u)),$$

which means that, if

$$E_i(u) = d \exp_{\mathbf{Y},x}(u) \left(\frac{\partial}{\partial u_i} \right)$$

and

$$E_i^j(u) = \sum_{m=0}^{j} \frac{(-1)^m}{(m+1)!} [(\text{ad}\,(u \cdot \mathbf{Y}))^m Y_i](\exp_{\mathbf{Y},x}(u)),$$

then

$$\|E_i(u) - E_i^j(u)\| = O(|u|^{j+1}).$$

Moreover, this estimate is uniform in x, as long as x stays in a compact set of V.

Proof Let f be a C^∞ function in the neighbourhood of x. By differentiating the Taylor series of $f(\exp_{\mathbf{Y},x}(u))$, one sees that the Taylor series of $\frac{\partial}{\partial u_1} f(\exp_{\mathbf{Y},x}(v))$ at 0 is

$$\sum_{m=0}^{+\infty} \frac{1}{(m+1)!} [\sum_{\ell=0}^{m} (u \cdot \mathbf{Y})^{m-\ell} Y_i (u \cdot \mathbf{Y})^\ell f](x).$$

In the algebra of differential operators on V, let R (resp. L) be right (resp. left) multiplication by $u \cdot \mathbf{Y}$ and $M = L - R = \text{ad}\,(u \cdot \mathbf{Y})$, so that

$$R^m = (L - M)^m = \sum_{p=0}^{m} (-1)^p C_m^p L^{m-p} M^p$$

and

$$\sum_{p=0}^{m} (u \cdot \mathbf{Y})^{m-\ell} Y_i (u \cdot \mathbf{Y})^\ell$$

$$= \sum_{\ell=0}^{m} L^{m-\ell} R^\ell (Y_i) = \sum_{\ell=0}^{m} \sum_{p=0}^{\ell} (-1)^p C_\ell^p L^{m-p} M^p (Y_i)$$

$$= \sum_{p=0}^{m} (-1)^p \left(\sum_{\ell=p}^{m} C_\ell^p \right) L^{m-p} M^p (Y_i) = \sum_{p=0}^{m} (-1)^p C_{m+1}^{p+1} M^p L^{m-p} (Y_i).$$

The Taylor series of $\frac{\partial}{\partial u_i} f(\exp_{\mathbf{Y},x}(u))$ thus reads

$$\sum_{m=0}^{+\infty} \sum_{p=0}^{m} \frac{(-1)^p}{(p+1)!} [\text{ad}\,(u \cdot \mathbf{Y})^p Y_i \frac{(u \cdot \mathbf{Y})^{m-p}}{(m-p)!} f](x),$$

that is to say

$$\left| \frac{\partial}{\partial u_i} f(\exp_{\mathbf{Y},x}(u)) - \sum_{p=0}^{j} \frac{(-1)^p}{(p+1)!} [\text{ad}\,(u \cdot \mathbf{Y})^p Y_i \sum_{m=0}^{j-p} \frac{(u \cdot \mathbf{Y})^m}{m!} f](x) \right|$$

$$= O(|u|^{j+1}).$$

We again use the fact that the Taylor series of $f(\exp_{\mathbf{Y},x}(u))$ is

$$\sum_{m=0}^{+\infty} \left[\frac{(u \cdot \mathbf{Y})^m}{m!} f\right](x),$$

to obtain

$$\left|\frac{\partial}{\partial u_i} f(\exp_{\mathbf{Y},x}(u)) - \sum_{p=0}^{j} \frac{(-1)^p}{(p+1)!}[\operatorname{ad}(u \cdot \mathbf{Y})^p Y_i f(\exp_{\mathbf{Y},x}(u))]\right| = O(|u|^{j+1}),$$

which is the claimed estimate.

We are now going to use the Campbell-Hausdorff formula to build another diffeomorphism from $\mathbb{R}^N$ to V. Let Q be the free associative algebra with unit generated by two non-commuting elements x and y. For $a, b \in Q$, $ab - ba$ is denoted by $[a, b]$. Let $\hat{Q}$ be the algebra of formal series in t with coefficients in Q. Then there exists a unique element $Z \in \hat{Q}$ such that $e^{tx}e^{ty} = e^Z$; Z is given by the Campbell-Hausdorff formula. In particular,

$$Z = (x + y)t + [x, y]\frac{t^2}{2} + R(x, y, t),$$

where $R(x, y, t)$ is a formal series in t only involving terms of degree at least three, the coefficient of t^m being a sum of brackets of order m in x and y.

It follows immediately that

$$e^{tx}e^{ty}e^{-tx}e^{-ty} = e^{[x,y]t^2 + R'(x,y,t)} \qquad (C.H.)$$

where $R'(x, y, t)$ has the same form as $R(x, y, t)$.

More generally, for $p \in \mathbb{N}^*$, let Q_p be the free associative algebra with unit and p generators $x_1, ..., x_p$, and $\hat{Q}_p$ the algebra of formal series in t with coefficients in Q_p. Let us define inductively, in $\hat{Q}_p$, $c_p(t)$ by $c_1(t) = e^{tx_p}$ and $c_i(t) = e^{tx_{p-i+1}}c_{i-1}(t)e^{-tx_{p-i+1}}c_{i-1}^{-1}(t)$; $c_p^{-1}(t)$ is the inverse of $c_p(t)$ in $\hat{Q}_p$. Applying (C.H.) several times, we obtain

$$c_p(t) = e^{[x_1,[x_2,...[x_{p-1},x_p]]...]t^p + R_p(x_1,...,x_p,t)} \qquad (C.H.)_p$$

hence

$$c_p^{-1}(t) = e^{-[x_1,[x_2,...[x_{p-1},x_p]]...]t^p - R_p(x_1,...,x_p,t)},$$

where $R_p(x_1, ..., x_p, t)$ is a formal series in t only involving terms of degree at least $p + 1$, the coefficient of t^m being a sum of brackets of order m in $x_1, ..., x_p$.

If $Y_1, ..., Y_p$ are C^∞ vector fields on the manifold V, let us now define $C_p(t, Y_1, ..., Y_p) = C_p(t)$ by $C_1(t) = \exp(tY_p)$ and

$$C_i(t) = C_{i-1}^{-1}(t)\exp(-tY_{p-i+1})C_{i-1}(t)\exp(tY_{p-i+1}).$$

$C_p(t)$ and its inverse $C_p^{-1}(t)$ are defined in the neighbourhood of every point of V for small enough t.

Let $x \in V$, f a C^∞ function defined in the neighbourhood of x, and $Z_1, ..., Z_q$ some C^∞ vector fields on V. Set

$$\widetilde{H}(t_1, ..., t_q) = f(\exp(t_1 Z_1) \cdots \exp(t_q Z_q)(x)),$$

and $H(t) = \widetilde{H}(t, ..., t)$. We have

$$\frac{\partial^{m_1}}{\partial t_1^{m_1}} \widetilde{H}(0, t_2, ..., t_q) = (Z_1^{m_1} f)(\exp(t_2 Z_2), ..., \exp(t_q Z_q)(x)),$$

and, iterating,

$$\frac{\partial^{m_1 + ... + m_q}}{\partial t_1^{m_1} \cdots \partial t_q^{m_q}} \widetilde{H}(0, ..., 0) = (Z_q^{m_q} \cdots Z_1^{m_1} f)(x).$$

The Taylor series of $\widetilde{H}$ at $(0, ..., 0)$ is thus $(e^{t_q Z_q} \cdots e^{t_1 Z_1} f)(x)$, and that of H at $(0, ..., 0)$ is $(e^{tZ_q} \cdots e^{tZ_1} f)(x)$.

In particular, if $\widetilde{c_p}(t)$ is the formal series obtained from substituting the Y_i for the indeterminates x_i in $c_p(t)$, the Taylor series at 0 of the function $f(\widetilde{c_p}(t))$ is $(\widetilde{c_p}(t)f)(x)$, that is to say, using $(C.H.)_p$,

$$\left(e^{[Y_1, [Y_2, ... [Y_{p-1}, Y_p]...]] t^p + R_p(Y_1, ..., Y_p, t)} f \right)(x),$$

which begins with

$$f(x) + (t^p [Y_1, [Y_2, ... [Y_{p-1}, Y_p]...]] f)(x) + ...$$

Likewise, the Taylor series at 0 of $f(c_p^{-1}(t))$ begins with

$$f(x) - t^p ([Y_1, [Y_2, ... [Y_{p-1}, Y_p]...]] f)(x) + ...$$

Now, if F_1 and F_2 are two C^∞ functions in the neighbourhood of 0, with Taylor series $F_1(t) = a + bt^p + ...$ and $F_2(t) = a - bt^p + ...$, it is easy to see that the function G defined by

$$G(t) = \begin{cases} F_1(t^{\frac{1}{p}}), & \text{if } t \geq 0 \\ F_2((-t)^{\frac{1}{p}}), & \text{if } t < 0, \end{cases}$$

is C^1 in the neighbourhood of 0, and satisfies $G'(0) = b$.

The function

$$\phi(t) = \begin{cases} C_p(t^{\frac{1}{p}}), & \text{if } t > 0 \\ C_p^{-1}((-t)^{\frac{1}{p}}) & \text{if } t < 0, \end{cases}$$

thus defines a C^1 class path from an interval containing 0 to V, reaching x at 0, and such that $\phi'(0) = [Y_1, [Y_2, ... [Y_{p-1}, Y_p]...]](x)$.

Let now $\mathbf{X} = \{X_1, ..., X_k\}$ be a Hörmander system of vector fields on V, $x \in V$, and $b = \{I_1, ..., I_N\} \in \mathcal{I}(k)^N$ such that $\{X_{I_1}(x), ..., X_{I_N}(x)\}$ is a basis of T_xV.

If $I = (i_1, ..., i_p) \in \mathcal{I}(k)$, put

$$\phi_I(t) = \begin{cases} C_p(t^{\frac{1}{p}}, X_{i_1}, ..., X_{i_p}) & \text{if } t \geq 0, \\ C_p^{-1}((-t)^{\frac{1}{p}}, X_{i_1}, ..., X_{i_p}) & \text{if } t < 0. \end{cases}$$

Finally, define ψ_b by $\psi_b(\theta) = \phi_{I_1}(\theta_1) \circ ... \circ \phi_{I_N}(\theta_N)(x)$ for $\theta = (\theta_1, ..., \theta_N)$. The above considerations show that the mapping ψ_b is C^1 from a neighbourhood of 0 in $\mathbb{R}^N$ to V, and satisfies

$$\frac{\partial}{\partial \theta_i} \psi_b(0) = X_{I_i}(x).$$

Hence it is a C^1-diffeomorphism from a neighbourhood of 0 in $\mathbb{R}^N$ to a neighbourhood of x in V. The local coordinates provided by ψ_b will be used in Chapter V and also in the next section.

III.4 Carnot–Carathéodory distances

Heuristically, the diffusion governed by the operator $\Delta = -\sum_{i=1}^k X_i^2$ "follows" the fields $X_1, ..., X_k$. Hence, the geometric objects associated in a natural way with Δ are the paths that stay tangent to the fields $X_1, ..., X_k$. If $\{X_1, ..., X_k\}$ is a Hörmander system, one can always connect two points of the manifold V by such a path. One defines in this way a distance which is suitable for the study of Δ, and is called the Carnot–Carathéodory distance.

More precisely, given $\mathbf{X} = \{X_1, ..., X_k\}$, a system of C^∞ vector fields on V, let $\mathcal{C}_{\mathbf{X}}$ be the set of all absolutely continuous paths $\gamma \colon [0, 1] \to V$, satisfying $\dot{\gamma}(t) = \sum_{i=1}^k a_i(t) X_i(\gamma(t))$, for almost every $t \in [0, 1]$.

Put

$$|\gamma| = \int_0^1 \left(\sum_{i=1}^k a_i^2(t) \right)^{\frac{1}{2}} dt,$$

and for $x, y \in V$,

$$\rho_{\mathbf{X}}(x, y) = \rho(x, y) = \inf\{|\gamma| \mid \gamma \in \mathcal{C}_X, \gamma(0) = x, \gamma(1) = y\}$$

if there exists at least one such path connecting x and y, and $+\infty$ otherwise.

It is easy to see that ρ is symmetric, satisfies the triangle inequality, and locally dominates the Euclidean distance induced by a chart.

Finally,

III.4.1 Proposition *If $\mathbf{X}$ is a Hörmander system, then $\rho(x, y) < +\infty$, for all $x, y \in V$, and ρ is a distance, which induces the topology of V.*

Proof By connectivity, it suffices to prove that $\rho(x, y) < +\infty$ for every $x \in V$, and every y in a neighbourhood of x.

Fix $x \in V$, and let $b = \{I_1, ..., I_N\} \in \mathcal{I}(k)^N$ be such that

$$\{X_{I_1}(x), ..., X_{I_N}(x)\}$$

is a basis of $T_x V$. Consider the mapping ψ_b constructed in Section III.3, and a neighbourhood U of 0 in $\mathbf{R}^N$ such that ψ_b is a C^1-diffeomorphism from U on the neighbourhood of x, $\Omega = \psi_b(U)$.

By developing all the factors that appear in the definition of ψ_b, we can write

$$\forall \theta \in U, \quad \psi_b(\theta) = \prod_{\alpha=1}^{M} \exp(\pm \eta_\alpha X_{i_\alpha})(x),$$

where $i_\alpha \in \{1, ..., k\}$, and $0 < \eta_\alpha \leq \|\theta\|^{\frac{1}{\delta}}$, where $\|.\|$ is the Euclidean norm and $\delta = \sup\{|I_j| \mid j \in \{1, ..., N\}\}$.

It follows easily from this expansion that for all $y \in \Omega$, $y = \psi_b(\theta)$ can be connected to x by a path belonging to $\mathcal{C}_\mathbf{X}$, of length smaller than $\|\theta\|^{\frac{1}{\delta}}$. We then have $\rho(x, y) < +\infty$, for all $y \in \Omega$ and more precisely

$$\rho(x, y) \leq C \rho_e^{1/\delta}(x, y),$$

where ρ_E is the Euclidean distance induced by the chart ψ_b^{-1}.

It is possible to show that we obtain a distance equivalent to ρ if we replace $\mathcal{C}_\mathbf{X}$ by $\widetilde{\mathcal{C}_\mathbf{X}}$, the set of continuous paths from $[0, 1]$ to V, such that $\dot{\gamma}(t) = \pm X_{j_i}(\gamma(t))$ on $]t_i, t_{i+1}[$, with $j_i \in \{1, ..., k\}$, where $t_0, ..., t_m$ is a subdivision of $[0, 1]$.

When $\mathbf{X} = \{X_1, ..., X_k\}$ is a Hörmander system of left invariant vector fields on a Lie group G, the distance $\rho = \rho_\mathbf{X}$ inherits the invariance property

$$\rho(gx, gy) = \rho(x, y), \quad \forall x, y, g \in G.$$

In particular $\rho(x, y) = \rho(e, x^{-1}y)$. We shall put $\rho(x) = \rho(e, x)$ so that $\rho(x, y) = \rho(x^{-1}y) = \rho(y^{-1}x)$.

It is clear that in general the distances $\rho_{\mathbf{X}_1}$ and $\rho_{\mathbf{X}_2}$ associated to two different Hörmander systems of vector fields $\mathbf{X}_1$ and $\mathbf{X}_2$ are not comparable. However, in the case of Lie groups, this distortion can only take place locally; indeed reasonable left invariant distances on a group are always equivalent at infinity.

More precisely, let G be a locally compact group. We shall say that a distance ρ on G is connected if it satisfies:

(C_1) ρ induces the topology of G;

(C_2) the closed ball for ρ centred at e of radius 1 is compact;

(C_3) there exists C such that, for all $x \in G$ satisfying $\rho(e, x) \geq 1$, there exist $\{x_0, ..., x_n\}$, a family of elements of G such that $x_0 = e$, $x_n = x$ and $\rho(x_i, x_{i+1}) \leq 1$, $i = 0, ..., n-1$, with $n \leq C\rho(e, x)$.

III.4.2 Proposition *Let ρ_1 and ρ_2 be two left invariant connected distances on G. Then there exists C such that $C^{-1}\rho_1(x) \leq \rho_2(x) \leq C\rho_1(x)$ as soon as $\rho_1(x) \geq 1$ or $\rho_2(x) \geq 1$.*

Proof Let $x \in G$ be such that $\rho_2(x) \geq 1$. We can find $x_0, ..., x_n$ such that $x_0 = e, x_n = x, \rho_2(x_i, x_{i+1}) \leq 1$, for all $i = 0, ..., n - 1$, and $n \leq C\rho_2(x)$, since ρ_2 is connected.

Let $a = \sup\{\rho_1(y) \mid \rho_2(y) \leq 1\}$ which is finite since $\{y \mid \rho_2(y) \leq 1\}$ is compact and ρ_1 continuous. Then

$$\rho_1(x) \leq \sum_{i=0}^{n-1} \rho_1(x_i, x_{i+1}) \leq na \leq Ca\rho_2(x).$$

We have used left invariance in the second inequality. Exchanging the rôles played by ρ_1 and ρ_2, we obtain the proposition.

One checks easily that the Carnot-Carathéodory distances are connected.

III.4.3 Remark The conditions (C_1) and (C_2) above can readily be replaced by the assumption that $\sup\{\rho_1(y) \mid \rho_2(y) \leq 1\}$ and $\sup\{\rho_2(y) \mid \rho_1(y) \leq 1\}$ are finite.

References and comments

In this chapter, we essentially adapt well-known standard results for further use. Section III.1 relies on the celebrated "sums of squares" theorem that Hörmander obtained in [65]. The proof given in [123] may be further adapted to check the uniform version III.1.5.

The first approach that we give to Theorem III.2.1 relies on the work of Bony ([15]); notice that the second approach clearly extends to general subelliptic, not necessarily sum of squares, operators (see [152]); the tools necessary for the construction of the hypoelliptic Dirichlet heat kernel can be found in [46] and [120]. Section III.3 is an exposition of classical facts about the exponential map on a manifold; it is very close to [64], Section I.6. Formula $(C.H.)_p$ is borrowed from [65]. The Carnot–Carathéodory distance of Section III.4 was introduced in [19]; the fundamental theorem III.4.1 is due to Chow ([22]). For more details on distances associated to vector fields, see [94].

CHAPTER IV

THE HEAT KERNEL ON NILPOTENT LIE GROUPS

At the end of this chapter, we shall know almost everything about the heat kernel and the Sobolev inequalities associated to a family of Hörmander fields on a nilpotent Lie group.

Later we shall obtain related results in much wider settings. Local questions will be studied on manifolds in Chapter V. Global questions will be studied on unimodular groups (Chapters VI, VII and VIII). We shall, in Chapter IX, even consider non-unimodular Lie groups as far as Sobolev inequalities are concerned.

But it is pleasant to see right away how the semigroup machinery of Chapter II and the considerations of Chapter III about sublaplacians yield complete results in the particular setting of nilpotent groups; this is because, in some sense, the geometry of these groups is not too complicated.

In Section 1 we recall general properties of nilpotent Lie groups, and in Section 2, we give examples. Section 3 is simple, but essential: we obtain for free a powerful scaled Harnack inequality. In Section 4, we show how to estimate the heat kernel with respect to the volume growth, thanks to this Harnack inequality. An analysis of the Lie algebra gives in Section 5 an estimate from above and below of the volume of the ball of radius t. In Sections 6 and 7, we draw fairly complete consequences of all this, using Chapter II; we also introduce a device which yields L^1 Sobolev inequalities and which we shall use again later.

IV.1 Some remarkable properties of nilpotent Lie groups

Let $\mathcal{L}$ be a finite dimensional Lie algebra, $\mathcal{L}_1 = \mathcal{L}$ and $\mathcal{L}_i = [\mathcal{L}, \mathcal{L}_{i-1}]$ for $i \geq 2$. The sequence $(\mathcal{L}_i)^*_{i \in \mathbf{N}}$ is a decreasing sequence of Lie sub-algebras of $\mathcal{L}$. The Lie algebra $\mathcal{L}$ is said to be *nilpotent* of *rank r* if $\mathcal{L}_{r+1} = \{0\}$. A Lie group G is said to be nilpotent of rank r if its Lie algebra is nilpotent of rank r. In all of this chapter, G is a connected nilpotent Lie group of rank r and $\mathcal{L} = \mathcal{L}_G$ is its Lie algebra, identified with the space of left invariant vector fields. We shall denote by exp the exponential mapping from $\mathcal{L}$ to G; in other words, X being a left invariant field, we shall denote by $\exp(X)$ the object which we denoted as $\exp(X)(e)$ in Chapter II, where e is the unit element of G.

Here are some classical facts which we shall use:

IV.1.1 *There exists a polynomial mapping P from $\mathcal{L} \times \mathcal{L}$ to $\mathcal{L}$ such that*

$$\exp(X)\exp(Y) = \exp P(X, Y), \quad \forall (X, Y) \in \mathcal{L} \times \mathcal{L}.$$

IV.1.2 *G is unimodular.*

IV.1.3   *The map* exp: $\mathcal{L} \to G$ *is onto. It is a diffeomorphism from a neighbourhood of zero in $\mathcal{L}$ to a neighbourhood of the unit element e in G which locally maps the Lebesgue measure on $\mathcal{L}$ to the Haar measure on G.*

IV.1.4   *If G is simply connected, the exponential map is a global diffeomorphism from $\mathcal{L}$ to G, and the Haar measure on G is the image of the Lebesgue measure on $\mathcal{L}$ under this diffeomorphism.*

IV.1.5   *If G is not simply connected, there exists a maximal compact subgroup K. K is different from $\{e\}$, contained in the centre of G, and G/K is a simply connected nilpotent group.*

We shall not give the proofs of these facts, but we shall outline some of them. For IV.1.1, let us recall that a polynomial mapping from one finite dimensional vector space into another is a mapping which, read in any pair of bases, has some polynomials as coordinate mappings; IV.1.1 follows thus immediately from the Campbell-Hausdorff formula and from the fact that G is nilpotent. The total degree of the map P is r.

It also follows from IV.1.1 that $\mathcal{L}$, endowed with the product P, is a Lie group, and that the exponential map is a homomorphism from $(\mathcal{L}, P)$ to G. The image of $\mathcal{L}$ under this homomorphism is a Lie subgroup of G which is open, since it contains a neighbourhood of the origin, and thus also closed. Since G is connected, the exponential map is surjective. If G is simply connected, the local diffeomorphism of Lie groups exp: $(\mathcal{L}, P) \to G$ admits a unique global extension, via one of Lie's fundamental theorems. Finally, we only have to explain the statements about measures; for IV.1.5, we refer the reader to the literature quoted in the References and Comments to this chapter.

Let us introduce on $\mathcal{L}$ a basis $(e_i)_{i=1}^n$ adapted to the filtration $\mathcal{L}_i, i \in \{1, ..., r\}$, i.e. such that, putting $n_i = n - \dim \mathcal{L}_i$ (so that $0 = n_1 < ... < n_{r+1} = n$), $(e_i)_{i \in \{n_j+1, ..., n\}}$ is, for every $j \in \{1, ..., r\}$, a basis of $\mathcal{L}_i$. It follows from this choice that the left multiplication by $g \in G$, denoted by L_g, is a polynomial map satisfying, for $g = (g_1, ..., g_n)$ and $x = (x_1, ..., x_n)$,

$$(L_g(x))_j = g_j + x_j + Q_g^j(x_1, ..., x_{j-1}),$$

where Q_g^j is a polynomial function of the first $j-1$ coordinates of x. The determinant of the mapping dL_g thus equals 1. The same properties obviously hold for the right multiplication by g, R_g and its differential dR_g. Hence the Lebesgue measure on $\mathcal{L}$, $dx = dx_1 \cdots dx_n$, pulled back to G by the map $\exp^{-1}$, is left and right invariant. This result is local in general and global if G is simply connected.

IV.2 Examples

Let us consider the product

$$(x, y, z)(x', y', z') = (x + x', y + y', z + z' + (xy' - x'y)/2)$$

on $\mathbb{R}^3$.

The group H obtained by endowing $\mathbb{R}^3$ with this product is a Lie group, and if we denote by X, Y, Z the left invariant fields whose values at the origin are $\frac{\partial}{\partial x}, \frac{\partial}{\partial y}, \frac{\partial}{\partial z}$, we have:

$$X(x,y,z) = \frac{\partial}{\partial x} - \frac{1}{2}y\frac{\partial}{\partial z},$$

$$Y(x,y,z) = \frac{\partial}{\partial x} + \frac{1}{2}x\frac{\partial}{\partial z},$$

$$Z(x,y,z) = \frac{\partial}{\partial z}.$$

X, Y, Z form a basis of the Lie algebra of H; the structure of the latter is particularly simple, since $[X,Y] = Z$ and the other brackets are zero. H is the Heisenberg group.

We may also realize H as the group of 3×3 upper triangular matrices whose diagonal coefficients equal 1. More generally, the group of $n \times n$ upper triangular matrices whose diagonal coefficients equal 1 gives an example of a nilpotent group of rank $n - 1$.

Let us come back to H and notice that it admits an adapted dilation group. Put

$$\phi_t \colon (x,y,z) \in H \mapsto \phi_t((x,y,z)) = (tx, ty, t^2 z), \quad t > 0.$$

It is obvious that

$$\phi_1 = \mathrm{Id}_H, \quad \phi_t \circ \phi_s = \phi_{ts}, \quad t, s > 0,$$

and also that

$$\forall (g,g') \in H^2, \forall t > 0, \quad \phi_t(gg') = \phi_t(g)\phi_t(g').$$

Finally the two generators X and Y of the Lie algebra of H are homogeneous of degree one with respect to ϕ_t; i.e. $X(f \circ \phi_t) = t(Xf) \circ \phi_t$, or $d\phi_t(X(x)) = tX(\phi_t(x))$.

A *stratified* group G is a simply connected nilpotent group whose Lie algebra $\mathcal{L}_G$ admits a direct sum decomposition

$$\mathcal{L}_G = \bigoplus_{i=1}^{r} V_i \quad \text{where } V_i = [V_1, V_{i-1}].$$

We say that V_i is the first *slice* in the stratification $\{V_i\}$ of $\mathcal{L}_G$; H is an example of a stratified group: $r = 2$, $V_1 = \mathrm{Vect}\{X,Y\}$, $V_2 = \mathrm{Vect}\{Z\}$.

Let G be a stratified group and $\{V_i\}$ a stratification of G. Let us consider, on the Lie algebra $\mathcal{L}_G$ of G, the family of dilations defined by

$$\forall t > 0, \forall X \in V_1, \quad \widetilde{\phi}_t(X) = tX,$$

which may be extended to all $\mathcal{L}_G$ by $\widetilde{\phi}_t(X) = t^i X$ if $X \in V_i$. The maps $\phi_t = \exp \circ \widetilde{\phi}_t \circ \exp^{-1} \colon G \to G$ satisfy:

$$\forall t, s > 0, \quad \phi_t \circ \phi_s = \phi_{ts}, \quad \phi_1 = \mathrm{Id}_G,$$
$$\forall t > 0, \forall (g, g') \in G^2, \quad \phi_t(gg') = \phi_t(g)\phi_t(g'),$$
$$\forall t > 0, \quad d\phi_t = \widetilde{\phi}_t.$$

In other words, G admits a group of dilations which is adapted to its structure. Conversely, we can easily show that if a simply connected Lie group G admits a dilation group $\{\phi_t \mid t > 0\}$ which is adapted to its structure, and if the space V_1 of the elements of $\mathcal{L}_G$ that are homogeneous of degree one with respect to ϕ_t generates $\mathcal{L}_G$, then G is nilpotent, in fact stratified, with V_1 as first slice.

Later on, we shall use the following abstract example. Let $\mathcal{L}(k, r)$ be the free nilpotent Lie algebra of rank r with k generators $e_1, ..., e_k$. By definition $\mathcal{L}(k, r)$ is the unique (up to isomorphism) nilpotent Lie algebra of rank r, such that for every nilpotent Lie algebra $\mathcal{L}$ of rank r and for every map α from $\{e_1, ..., e_k\}$ to $\mathcal{L}$, there exists a unique homomorphism $\widetilde{\alpha}$ from $\mathcal{L}(k, r)$ to $\mathcal{L}$ which extends α. We can construct $\mathcal{L}(k, r)$ by starting from the free Lie algebra with k generators, $\mathcal{L}(k)$ (whose construction is classical), and putting $\mathcal{L}_1(k) = \mathcal{L}(k)$, $\mathcal{L}_i(k) = [\mathcal{L}(k), \mathcal{L}_{i-1}(k)]$ for $i \geq 2$, and $\mathcal{L}(k, r) = \mathcal{L}(k)/\mathcal{L}_{r+1}(k)$. Notice that $\mathcal{L}(k, r)$, unlike $\mathcal{L}(k)$, is finite dimensional.

We shall denote by $\mathcal{N}(k, r)$ the free nilpotent group of rank r, with k generators, which is the simply connected group of Lie algebra $\mathcal{L}(k, r)$; $\mathcal{N}(k, r)$ is an example of a stratified group for which we can construct a dilation group as above by taking $V_1 = \mathrm{Vect}\{e_1, ..., e_k\}$.

IV.3 Harnack inequalities for nilpotent Lie groups

This section is devoted to the faithful generalization of the classical parabolic Harnack theorem in Euclidean space to the setting of nilpotent Lie groups. Given $\mathbf{X} = \{X_1, ..., X_k\} \subset \mathcal{L}_G$ a Hörmander system of left invariant vector fields on a nilpotent group G, let $B(x, r)$, $x \in G$, $r > 0$, be the ball centred at x of radius r for the distance $\rho = \rho_{\mathbf{X}}$ associated with $\mathbf{X}$. If $I = (i_1, ..., i_\alpha) \in \mathcal{I}(k)$, we shall denote by X^I the left invariant differential operator defined as

$$X^I f = X_{i_1} \cdots X_{i_\alpha} f,$$

which must not be confused with

$$X_I = [X_{i_1}, [X_{i_2}, ..., X_{i_\alpha}]...].$$

One denotes by $\Delta_{\mathbf{X}} = \Delta = -\sum_{i=1}^{k} X_i^2$ the sublaplacian associated with $\mathbf{X}$ and $\nabla f = (X_1 f, ..., X_k f)$.

IV.3.1 Theorem *Let G be a nilpotent group, $\mathbf{X} = \{X_1, ..., X_k\} \subset \mathcal{L}_G$ a Hörmander system. Given $0 < \alpha < \beta < 1$, $0 < \delta < 1$, $m \in \mathbf{N}$, $I \in \mathcal{I}(k)$,* there exists a constant C such that for every $x \in G$, every $R > 0$ and every positive solution u of $(\frac{\partial}{\partial t} + \Delta)u = 0$ in $]0, R[\times B(x, \sqrt{R})$, we have

$$\sup_{y \in B(x, \delta\sqrt{R})} \left| X^I \left(\frac{\partial}{\partial t}\right)^m u(\alpha R, y) \right| \leq C R^{-m-|I|/2} \inf_{y \in B(x, \delta\sqrt{R})} u(\beta R, y).$$

We start with the particular case of stratified groups.

IV.3.2 Proposition   *Theorem IV.3.1 holds under the additional hypothesis that G is stratified and that $\mathbf{X} = \{X_1, ..., X_k\}$ is included in the first slice of some stratification of $\mathcal{L}_G$.*

Proof   Under this hypothesis, there exists a group $\{\phi_t \mid t > 0\}$ of dilations which is adapted to the structure of G and such that

$$d\phi_t(X_i) = tX_i, \quad t > 0, i \in \{1, ..., k\}.$$

Consider now a positive solution u of $(\frac{\partial}{\partial t} + \Delta)u = 0$ in $]0, R[\times B(x, \sqrt{R})$, and define

$$v_u:]0, 1[\times B(e, 1) \to \mathbb{R}^+$$

by $v_u(t, \xi) = u(Rt, x\phi_{\sqrt{R}}(\xi))$. We check at once that $v = v_u$ is a positive solution of $(\frac{\partial}{\partial t} + \Delta)v = 0$ in $]0, 1[\times B(e, 1)$ and Theorem III.2.1 yields a constant C independent of v (that is, from x, R and u) such that

$$\inf_{\xi \in B(e, \delta)} \left| X^I \left(\frac{\partial}{\partial t}\right)^m u(\alpha, \xi) \right| \leq C \inf_{\xi \in B(e, \delta)} u(\beta, \xi).$$

Furthermore,

$$X^I \left(\frac{\partial}{\partial t}\right)^m u(\alpha, \xi) = R^{m+|I|/2} [X^I \left(\frac{\partial}{\partial t}\right)^m u](\alpha R, x\phi_{\sqrt{R}}(\xi))$$

and

$$\phi_{\sqrt{R}}(B(e, \delta)) = B(e, \delta\sqrt{R}).$$

This proves Proposition IV.3.2.

Proof of Theorem IV.3.1   In a very remarkable way, the proof of Theorem IV.3.1 essentially reduces to that of Proposition IV.3.2.

Indeed, let r be the rank of the nilpotent group G. By the definition of the free nilpotent algebra $\mathcal{L}(k, r)$ introduced in Section IV.2, there exists a Lie algebra homomorphism $\tilde{\Pi}$ from $\mathcal{L}(k, r)$ to $\mathcal{L}_G$ such that $\tilde{\Pi}(e_i) = X_i$, $i \in \{1, ..., k\}$; since $\mathbf{X} = \{X_1, ..., X_k\}$ is a Hörmander system, $\tilde{\Pi}$ is onto. Through the exponential maps, $\tilde{\Pi}$ induces a surjective homomorphism π from $\mathcal{N}(k, r)$ to G: $\pi = \exp_G \circ \tilde{\Pi} \circ \exp_{\mathcal{N}}^{-1}$; we have of course $d\pi = \tilde{\Pi}$. We easily check that, if $\tilde{B}(\tilde{x}, t)$ is the ball centred at $\tilde{x} \in \mathcal{N}(k, r)$ and of radius

$t > 0$ for the distance associated to the system $e = \{e_1, ..., e_k\}$ on $\mathcal{N}(k, r)$, then

$$\pi(\tilde{B}(\tilde{x}, t)) = B(\pi(\tilde{x}), t).$$

Consider now a positive solution u of $(\frac{\partial}{\partial t} + \Delta)u = 0$ in $]0, R[\times B(x, \sqrt{R}) \subset \mathbb{R}^{+*} \times G$, and set

$$v_u = u \circ \pi \colon \mathcal{N}(k, r) \to \mathbb{R}^+.$$

Put $\tilde{\Delta} = -\sum_{i=1}^{k} e_i^2$; the function v_u is obviously a positive solution of

$$(\frac{\partial}{\partial t} + \tilde{\Delta})v = 0$$

in $]0, R[\times \tilde{B}(\tilde{x}, \sqrt{R}) \subset \mathbb{R}^{+*} \times \mathcal{N}(k, r)$, for every $\tilde{x} \in \mathcal{N}(k, r)$ such that $\Pi(\tilde{x}) = x$. Moreover, for all $I \in \mathcal{I}(k)$, $m \in \mathbb{N}$,

$$[e^I \left(\frac{\partial}{\partial t}\right)^m v_u](t, \tilde{x}) = [X^I \left(\frac{\partial}{\partial t}\right)^m u](t, \Pi(\tilde{x})).$$

It follows that Proposition IV.3.2, applied to $\mathcal{N}(k, r)$, $e = \{e_1, ..., e_k\}$ and the function v_u, immediately yields the inequality of Theorem IV.3.1 for u.

To obtain Theorem IV.3.1, we only used the local Harnack theorem III.2.1, and not the sophisticated version stated in Theorem III.2.4. But this has been done thanks to algebraic constructions that are entirely specific to the setting of nilpotent Lie groups.

We are now going to prove, using Theorem IV.3.1 and by a classical method, a global result for positive solutions of $(\frac{\partial}{\partial t} + \Delta)u = 0$ in $\mathbb{R}^{+*} \times G$.

IV.3.3 Theorem *Let G and $\mathbf{X}$ be as in IV.3.1. There exists $C > 0$ such that every positive solution u of $(\frac{\partial}{\partial t} + \Delta)u = 0$ in $\mathbb{R}^{+*} \times G$ satisfies*

$$\forall x, y \in G, \quad \forall t > 0, \quad u(t, y) \leq Cu(2t, x) \exp(C\rho^2(x, y)/t),$$

where ρ is the distance associated with $\mathbf{X}$.

Proof Theorem IV.3.1 gives in particular

$$u(s, x) \leq Cu(2s, y), \quad s > 0, \quad \rho(x, y) \leq 2\sqrt{s}.$$

Let us apply this result to the function $(s, x) \mapsto u(t' + s, x)$ for fixed t'. This gives $u(t' + s, x) \leq Cu(t' + 2s, y)$, $s > 0$, $t' \geq 0$, $\rho(x, y) \leq 2\sqrt{s}$, and, by choosing $t' = t - s$,

$$u(t, x) \leq Cu(t + s, y), \quad t \geq s > 0, \quad \rho(x, y) \leq 2\sqrt{s}. \qquad (1)$$

If $\rho(x, y) \leq 2\sqrt{t}$, the inequality to prove reduces to (1) with $s = t$. Assume now $\rho(x, y) > 2\sqrt{t}$; let m be the integer part of $4\rho^2(x, y)/t$, and let $s = t/m$. There certainly exists a sequence $x_0 = x, x_1, ..., x_m = y$ of points in G, such that $\rho(x_i, x_{i+1}) \leq 2\sqrt{s}$ and $m\sqrt{s} \leq 2\rho(x,y)$. From the inequality (1), we deduce that for $i = 0, ..., m-1$,

$$u(t + is, x_i) \leq Cu(t + (i+1)s, x_{i+1}),$$

and finally

$$u(t, x) \leq C^m u(2t, y).$$

Taking into account the value of m, we obtain the conclusion of Theorem IV.3.3.

IV.4 Estimates of the heat kernel

Let G be a nilpotent Lie group and $\mathbf{X} = \{X_1, ..., X_k\}$ a Hörmander system on G. Consider, as in II.5.1, the symmetric markovian semigroup $H_t = e^{-t\Delta}$, $t \geq 0$, associated with $\Delta = -\sum_{i=1}^k X_i^2$. The left invariance of Δ shows that H_t is given by a right convolution:

$$H_t f(x) = \int_G f(y) h_t(y^{-1}x) \, dy.$$

Let us recall that $(t, x) \mapsto h_t(x)$ is a positive solution of $(\frac{\partial}{\partial t} + \Delta)u = 0$ and thus, by hypoellipticity, a C^∞ function on $\mathbb{R}^{+*} \times G$, and that $\|h_t\|_1 = 1$.

One can apply Theorem IV.3.1 to $u(t, x) = h(t, x)$, which gives in particular

$$h_t(e) \leq C \inf_{B(e,\sqrt{t})} h_{2t}(y), \quad \forall t > 0,$$

hence

$$V(\sqrt{t}) h_t(e) \leq C \int_{B(e,\sqrt{t})} h_{2t}(y) \, dy \leq C \int_G h_{2t}(y) \, dy \leq C.$$

Here $V(t)$ is the Haar measure of the ball $B(x, t)$, for any $x \in G$. This gives

IV.4.1 Proposition *Let G be a nilpotent Lie group, $\mathbf{X}$ a Hörmander system and h_t the associated heat kernel. Then $h_t(e) \leq CV(\sqrt{t})^{-1}$, for all $t > 0$.*

In fact, with a little more work, we are going to obtain the much better

IV.4.2 Theorem *Let G be a nilpotent Lie group, $\mathbf{X}$ a Hörmander system and h_t the associated heat kernel. Then for all $m \in \mathbb{N}$, $I \in \mathcal{I}(k)$, $\varepsilon > 0$, there exists C such that*

$$\left| \left(\frac{\partial}{\partial t} \right)^m X^I h_t(x) \right| \leq C t^{-m-|I|/2} V(\sqrt{t})^{-1} \exp(-\rho^2(x)/4(1+\varepsilon)t),$$

for all $x \in G$ and $t > 0$.

Proof Notice first that the result for $m \in \mathbb{N}$ and $I \in \mathcal{I}(k)$ follows immediately from the particular case $m = 0$, $I = \emptyset$ thanks to Theorem IV.3.1. So suppose $m = 0$, $I = \varnothing$. Let $\phi \in C_0^\infty(G)$ satisfy $|\nabla\phi| \leq 1$ and let $\lambda \in \mathbb{R}$. Consider the operator $Bf = e^{-\lambda\phi}\Delta(e^{\lambda\phi}f)$, $f \in C_0^\infty(G)$. One has

$$(Bf, f) = (\nabla e^{\lambda\phi}f, \nabla e^{-\lambda\phi}f)$$
$$= \|\nabla f\|_2^2 - \lambda^2 \int |\nabla\phi|^2|f|^2 + \lambda \int f\nabla\phi \cdot \nabla\overline{f} + \overline{f}\nabla\phi \cdot \nabla f.$$

Hence, because of the hypothesis on ϕ,

$$\mathrm{Re}\,(Bf, f) \geq \|\nabla f\|_2^2 - \lambda^2\|f\|_2^2.$$

In particular the semigroup $e^{-tB} = e^{-\lambda\phi}H_t(e^{\lambda\phi})$ satisfies $\|e^{-tB}\|_{2\to2} \leq e^{\lambda^2 t}$, $t \geq 0$. If we test this inequality on the characteristic functions $1_{B(e,\sqrt{t})}$ and $1_{B(x,\sqrt{t})}$, we get

$$\int_{\xi\in B(x,\sqrt{t})}\int_{\zeta\in B(e,\sqrt{t})} h_t(\xi^{-1}\zeta)e^{\lambda(\phi(\xi)-\phi(\zeta))}\,d\xi\,d\zeta$$
$$\leq Ce^{\lambda^2 t}V(\sqrt{t}), \quad x \in G, t > 0.$$

Using Theorem IV.3.1 and the hypothesis on ϕ again, we obtain

$$h_{(1-\varepsilon)t}(x) \leq C_\varepsilon V(\sqrt{t})^{-1}\exp(\lambda^2 t + \lambda(\phi(x) - \phi(e)) + 2|\lambda|\sqrt{t}).$$

Let us fix x and t and choose for ϕ a compactly supported approximation of $\zeta \mapsto \rho(e,\zeta)$, such that $\phi(e) \simeq 0$, $\phi(x) \simeq \rho(e,x) = \rho(x)$; finally let us take $\lambda = -\rho(x)/2t$. This yields

$$h_{(1-\varepsilon)t}(x) \leq C_\varepsilon V(\sqrt{t})^{-1}\exp(-\rho^2(x)/4t + \rho(x)/\sqrt{t}),$$

and by changing $(1 - \varepsilon)t$ into t,

$$h_t(x) \leq C_\varepsilon V(\sqrt{t})^{-1}\exp(-\rho^2(x)/4(1 + \varepsilon)t)$$

which is the expected conclusion.

Our next result is the companion lower bound for the heat kernel h_t.

IV.4.3 Theorem *Let G be a nilpotent Lie group, $\mathbf{X}$ a Hörmander system on G. There exist $C, C' > 0$ such that the heat kernel satisfies*

$$h_t(x) \geq CV(\sqrt{t})^{-1}\exp(-C'\rho^2(x)/t),$$

for all $x \in G$ and $t > 0$.

Proof Theorem IV.3.3 shows that

$$h_t(x) \geq Ch_{t/2}(e)\exp(-C'\rho^2(x)/t),$$

for all $x \in G$ and $t > 0$. Now $h_t(e) = \|H_t\|_{1\to\infty}$ is a non-increasing function of t, and $h_{t/2}(e) \geq h_t(e)$. This reduces the proof of the theorem to the proof of the central estimate $h_t(e) \geq C'V(\sqrt{t})^{-1}$, $\forall t > 0$.

To prove this estimate, fix $t > 0$ and consider the function u in $\mathbb{R} \times G$ defined by

$$u(s, x) = \begin{cases} H_s 1_{B(\sqrt{t})}(x) & \text{for } s > 0, \\ u(s, x) = 1 & \text{for } s \leq 0. \end{cases}$$

One checks easily that u is a positive solution of $(\frac{\partial}{\partial t} + \Delta)u = 0$ in $\mathbb{R} \times B(\sqrt{t})$ (the only problem is at $s = 0$, but there, for $x \in B(\sqrt{t})$, u is continuous, and classical arguments show that this is enough). Now, Theorem IV.3.1 yields a constant C such that

$$u(0, e) \leq Cu(t, e), \quad \forall t > 0.$$

But $u(0, e) = 1$ and

$$u(t, e) = \int_{B(\sqrt{t})} h_t(y) dy \leq h_t(e) V(\sqrt{t}).$$

This proves the central estimate, and the theorem.

The above results invite us to estimate now the function $V(t)$. To get an idea of what we are going to obtain, let us suppose for a moment that G is stratified and that $\mathbf{X}$ is included in the first slice. Consider then the dilation group ϕ_t, $t > 0$ on G. We see immediately that for every measurable set $|\phi_t(\Omega)| = t^D \phi(\Omega)$, $t > 0$, where $D = \sum_{i=1}^{r} i \dim V_i$, with the notation of Section 2. In particular, $B(e, t) = \phi_t(B(e, 1))$ and thus $V(t) = Ct^D$. Thus Theorems IV.4.2 and IV.4.3 give in this case $\forall x \in G, t > 0$,

$$Ct^{-D/2} \exp(-C' \rho^2(x)/t) \leq h_t(x) \leq C_\varepsilon t^{-D/2} \exp(-\rho^2(x)/4(1 + \varepsilon)t).$$

In the case of a general nilpotent Lie group G endowed with a Hörmander system $\mathbf{X}$, we are going to see that $V(t) \simeq t^d$, for small t, and $V(t) \simeq t^D$, for large t, where D is an invariant of G and d depends on G and $\mathbf{X}$.

IV.5 Estimates of the volume

To estimate the function $V(t) = \xi(B(e, t))$, we are essentially going to rely on the facts IV.1.1 and IV.1.5 recalled at the beginning of this chapter, and on a systematic use of the Campbell-Hausdorff formula. For a moment, let us use the notation $[x, y] = xyx^{-1}y^{-1}$ for $(x, y) \in G \times G$. In Section III.3 we noticed that in full generality,

$$[\exp Y_1, [\exp Y_2, ..., \exp Y_N]...] = \exp([Y_1, [Y_2, ..., Y_N]...] + R_N) \qquad (1)$$

where R_N is a sum of multiples of brackets of the $Y_1,...,Y_N$ of order greater than or equal to $N + 1$. But if G is nilpotent of rank r, R_r vanishes, so that

$$[\exp Y_1, [\exp Y_2, ..., \exp Y_r]...] = \exp([Y_1, [Y_2, ..., Y_r]...]).$$

Much more generally, we have:

IV.5.1 Lemma *Given a sequence $I_1, ..., I_m$ of multi-indices belonging to $\mathcal{I}(k)$, there exists an integer M and two sequences $(\alpha_1, ..., \alpha_M) \in \mathbb{R}^M$, $(j_1, ..., j_n) \in \mathcal{I}(k)$ such that*

$$\exp\left(\sum_{s=1}^{m} Y_{I_s}\right) = \prod_{i=1}^{M} \exp(\alpha_i Y_{j_i}).$$

Proof We proceed by backward iteration on $\nu = \min\{|I_s| \mid s \in \{1, ..., m\}\}$. Indeed, if $\nu = r$, the result reduces to our remark just before the lemma since, by Campbell-Hausdorff,

$$\exp\left(\sum_{s=1}^{M} Y_{I_s}\right) = \prod_{s=1}^{M} \exp(Y_{I_s}).$$

Now, if $\min\{|I_s| \mid s \in \{1, ..., n\}\} = \alpha < r$, suppose that $|I_1| = |I_2| = ... = |I_\ell| = \alpha$ and $|I_s| \geq \alpha + 1$ for $s > \ell$. Then, using (1) and the Campbell-Hausdorff formula, we easily construct M, $(\varepsilon_i)_{i=1}^{M}$ (where $\varepsilon_i = \pm 1$) and $(j_i)_{i=1}^{M} \in \mathcal{I}(k)$ such that

$$\prod_{i=1}^{M} \exp(\varepsilon_i Y_{j_i}) \exp\left(\sum_{s=1}^{m} Y_{I_s}\right) = \exp\left(\sum_{s=\ell+1}^{m} Y_{I_s} + R\right)$$

where R is a sum (which is finite since G is nilpotent) of multiples of brackets of the fields Y_i, $i \in \{1, ..., k\}$ of order greater than or equal to $\alpha + 1$. We can thus obtain the proof of IV.5.1 by induction.

Given the Hörmander system $\mathbf{X} = \{X_1, ..., X_k\}$, consider the set $\mathcal{B}$ of all subsets $b \subset \mathcal{I}(k)$, $b = \{I_1, ..., I_n\}$ such that $\{X_{I_1}, ..., X_{I_n}\}$ is a basis of $\mathcal{L}$. Since G is nilpotent, $\mathcal{B}$ is a finite set. Let us put, for $b \in \mathcal{B}$,

$$\omega(b) = \sum_{j=1}^{n} |I_j| \quad \text{if } b = \{I_1, ..., I_n\}.$$

In a moment, we are going to show that $V(t) \simeq \sum_{b \in \mathcal{B}} t^{\omega(b)}$, $t > 0$, if G is simply connected. To obtain a more striking result it suffices to determine $\sup\{\omega(b) \mid b \in \mathcal{B}\}$ and $\inf\{\omega(b) \mid b \in \mathcal{B}\}$. To find $\sup\{\omega(b) \mid b \in \mathcal{B}\}$, we only have to construct a basis b^+ by choosing first the greatest possible number of free vectors in $\mathcal{L}_r$, then completing step by step to obtain successively bases for $\mathcal{L}_{r-1}, \mathcal{L}_{r-2}, ..., \mathcal{L}_1 = \mathcal{L}$ (cf. Remark IV.5.9). We thus have

$$\sup\{\omega(b) \mid b \in \mathcal{B}\} = \omega(b^+) = \sum_{i=1}^{r} i \dim [\mathcal{L}_i/\mathcal{L}_{i+1}].$$

To find $\inf\{\omega(b) \mid b \in \mathcal{B}\}$, we only have to choose a basis b^- by choosing first the greatest number of free vectors in

$$K_X^1 = \text{Vect}\{X_1, ..., X_k\},$$

then to complete step by step to obtain successively bases for $K_X^2, ..., K_X^s = \mathcal{L}$, where K_X^α is the linear span of the brackets of $X_1, ..., X_k$ of length smaller than or equal to α (cf. again Remark IV.5.9). We get

$$\inf\{\omega(b) \mid b \in \mathcal{B}\} = \omega(b^-) = \sum_{i=1}^{s} i \dim(K_X^i/K_X^{i-1}).$$

Finally $V(t) \simeq \sum_{b \in \mathcal{B}} t^{\omega(b)}$ gives

$$V(t) \simeq \begin{cases} t^{\omega(b^-)} & \text{if } 0 < t \leq 1, \text{ and} \\ t^{\omega(b^+)} & \text{if } t \geq 1. \end{cases}$$

This justifies the following definitions:

IV.5.2 Definition   Given a connected Lie group G, and a Hörmander system $\mathbf{X} \subset \mathcal{L}_G$ on G, the *local dimension of* $(G, \mathbf{X})$ is the integer

$$d(G, \mathbf{X}) = d = \sum_{i=1}^{+\infty} i \dim[K_X^i/K_X^{i-1}].$$

IV.5.3 Definition   Given a simply connected nilpotent group, the *dimension at infinity of* G is the integer $D(G) = D = \sum_{i=1}^{r} i \dim[\mathcal{L}_i/\mathcal{L}_{i+1}]$.

IV.5.4 Definition   Given a nilpotent group G, let K be the unique maximal compact subgroup of G, $G' = G/K$ and $\mathcal{L}'$ the Lie algebra of G'. The *dimension at infinity* of G is the integer $D(G) = D(G') = D = \sum_{i=1}^{+\infty} i \dim[\mathcal{L}'_i/\mathcal{L}'_{i+1}]$.

Concerning this last definition, the reader is referred to IV.1.5. Notice in particular that IV.5.4 and IV.5.3 coincide if G is simply connected, since then $K = \{e\}$.

Definition IV.5.4 is justified by

IV.5.5 Lemma   *If G, K and G' are as in IV.5.4, we have $V_G(t) \simeq V_{G'}(t)$ for $t \geq 1$ (whatever the Hörmander systems considered on G and G' may be).*

Proof   Let π be the canonical projection from G onto G'. Since K is normal (and even central), the image under π of the Haar measure ξ on G is a Haar measure ξ' on G'. Let t be large enough to ensure $K \subset B(e, t)$, where $B(e, t)$ is the ball associated with some Hörmander system $\mathbf{X}$ on G. It is clear that $\pi(B(e, t)) = B'(e', t)$, where e' is the origin in G' and B' the ball in G' associated with the system $d\pi(\mathbf{X})$. We then have $B(e, t) \subset KB(e, t) \subset B^2(e, t) = B(e, 2t)$, hence $\xi(B(e, t)) \leq \xi(KB(e, t)) = \xi'(B'(e', t)) \leq \xi(B(e, 2t))$. Together with the fact that all connected distances on G and G' respectively are equivalent (see III.4.2), this ends the proof of the lemma.

Let us now attack the lower estimate of $V(t)$ for fixed G and $\mathbf{X}$.

IV.5.6 Proposition *Let G be a nilpotent group and $\mathbf{X}$ a Hörmander system on G. Then, for every $b \in \mathcal{B}$:*

(i) *if G is simply connected,*

$$V(t) \geq Ct^{\omega(b)}, \quad \forall t \geq 0;$$

(ii) *for any G,*

$$V(t) \geq Ct^{\omega(b)}, \quad \forall t, 0 < t \leq 1.$$

Finally, in any case, if d and D are, respectively, the local dimension and the dimension at infinity of G, X, we have

$$V(t) \geq Ct^d \text{ if } 0 \leq t \leq 1 \text{ and } V(t) \geq Ct^D \text{ if } t \geq 1.$$

Proof We first concentrate on (i). Let $b \in \{I_1, ..., I_n\} \in \mathcal{B}$. Let $R_b(t)$ be the parallelepiped of $\mathbb{R}^n = \mathcal{L}$ built on the basis $\{X_{I_1}, ..., X_{I_n}\}$ and defined by

$$R(t) = R_b(t) = \{Y = \sum_{\ell=1}^{n} y_\ell X_{I_\ell} \mid |y_\ell| \leq t^{|I_\ell|}\}.$$

We obviously have $\mathrm{Vol}_{\mathbb{R}^n}(R(t)) = Ct^{\omega(b)}$, $t \geq 0$, since $\omega(b) = \sum_{\ell=1}^{n}|I_\ell|$. Moreover, since G is simply connected, the image of the Lebesgue measure of $\mathcal{L} = \mathbb{R}^n$ under the exponential map is the Haar measure on G. It follows that, if $\phi(y) = \exp(\sum_{\ell=1}^{n} y_\ell X_{I_\ell})$, $y = (y_1, ..., y_n)$, we have

$$\xi(\phi(R(t))) = Ct^{\omega(b)}, \quad t \geq 0.$$

We now only have to show that $\phi(R(t)) \subset B(ct)$ for a constant c independent of $t > 0$. This is an immediate consequence of Lemma IV.5.1. Applying this lemma twice, we get first for $(y_1, ..., y_n) \in R(t)$

$$\phi(y) = \prod_{i=1}^{m} \exp(\alpha_j y_{j_i} Y_{j_i}),$$

where $Y_i = X_{I_i}$, $i \in \{1, ..., n\}$; then for each term of this product

$$\exp(\alpha y Y) = \prod_{i=1}^{m} \exp(\beta_i |\alpha y|^{\frac{1}{q}} X_{j_i}),$$

where $y = |I_\ell|$ if $Y = Y_\ell = X_{I_\ell}$ and thus $|y|^{\frac{1}{q}} \leq t$; take care, here y denotes one of the y_ℓ ! In other words, $\rho(e, \phi(y)) = \rho(\phi(y)) \leq Ct$, which ends the proof of (i).

The proof of (ii) is the same, but without the hypothesis of simple connectivity; the mapping of the volume can only take place on a neighbourhood of 0 in $\mathbb{R}^n$ and $\mathcal{L}$, and the result only holds for t small. The final conclusion follows from Definitions IV.5.2, IV.5.3, IV.5.4, from the remarks just before them, from Lemma IV.5.5 and from (i) and (ii).

We now turn to the estimate from above of the function $V(t)$.

IV.5.7 Proposition   *Let G be a nilpotent group, $\mathbf{X}$ a Hörmander system on G. Then*

(i) *if G is simply connected, $V(t) \leq C \sum_{b \in \mathcal{B}} t^{\omega(b)}$, $t \geq 0$,*

(ii) *for any G, $V(t) \leq C \sum_{b \in \mathcal{B}} t^{\omega(b)}$, $0 \leq t \leq 1$.*

Finally, in any case, if d and D are the local dimension and the dimension at infinity of G, X, we have:

$$V(t) \leq Ct^d \ if \ 0 \leq t \leq 1, \ and \ V(t) \leq Ct^D \ if \ t \geq 1.$$

Proof It relies on the Campbell-Hausdorff formula. According to Section III.4, there exists a constant M such that, if $x \in B(t)$, we can find a continuous path $\gamma \colon [0, T] \to G$, where $T \leq Mt$, such that $\gamma(0) = e$, $\gamma(T) = x$ and that there exists a partition $0 = s_0, ..., s_N = T$ such that $\frac{d}{ds}\gamma(s) = \pm X_{i_j}$ if $s \in {]}s_{i-1}, s_i{[}$. Clearly we can suppose that this partition is regular, so that

$$x = \prod_{i=1}^{N} \exp\left(\pm(T/N)X_{j_i}\right). \tag{2}$$

Applying again and again the Campbell-Hausdorff formula, putting $Y_i = X_{j_i}$, $i \in \{1, ..., M\}$ and using Jacobi's identity, we obtain

$$x = \exp\left(\sum_{\substack{I \in \mathcal{I}(N) \\ |I| \leq r}} \gamma_{N,I}(T/N)^{|I|}Y_I \right) \tag{3}$$

where the $\gamma_{N,I}$'s only depend on the distribution of $+$ and $-$ in (2) and on N, and satisfy $|\gamma_{N,I}| \leq \gamma$ where γ does not depend on N (because G is nilpotent). Let us now consider the collection of all X_I's, $I \in \mathcal{I}(k)$, $|I| \leq r$. Each Y_I, $I \in \mathcal{I}(k)$, can be written X_J, where $J \in \mathcal{I}(k)$ and $|I| = |J|$. Moreover, the number of Y_I with $|I| = \alpha$ is smaller than N^α. Thus (3) yields:

$$x = \exp\left(\sum_{\substack{I \in \mathcal{I}(N) \\ |I| \leq r}} a_{N,I}T^{|I|}X_I \right),$$

where $|a_{N,I}| \leq \gamma$. It follows, in the previous notation, that

$$x \in \exp\left(\bigcup_{b \in \mathcal{B}} R_b(\gamma T) \right)$$

and thus

$$B(t) \subset \exp[\bigcup_{b \in B} R_b(Mt)], \quad t \geq 0.$$

Since the map exp preserves measures under hypothesis (i), we obtain the stated result. In case (ii) this again holds for t small; the final conclusion follows as in IV.5.6.

We can thus state

IV.5.8 Theorem *Given a nilpotent group G whose dimension at infinity is D and $\mathbf{X}$ a Hörmander system such that the local dimension of $(G, \mathbf{X})$ is d, then $V(t) \simeq \sum_{b \in \mathcal{B}} t^{\omega(b)}$, i.e. $V(t) \simeq t^d$ if $0 \leq t \leq 1$ and $V(t) \simeq t^D$ if $t \geq 1$.*

IV.5.9 Remark Suppose G simply connected and consider the two filtrations of $\mathcal{L}$:

$$\mathcal{L} = \mathcal{L}_1 \supset \mathcal{L}_2 \supset ... \supset \mathcal{L}_r$$

and

$$K_X^0 = \{0\} \subset K_X^1 \subset ... \subset K_X^s = \mathcal{L}.$$

Let us derive from each one of them a decomposition $\mathcal{L} = \oplus_{i=1}^r V_i$ (resp. $\mathcal{L} = \oplus_{i=1}^s U_i$) such that $\mathcal{L}_j = \oplus_{i=j}^r V_i$ (resp. $K_X^j = \oplus_{i=1}^j U_i$). In this notation, we have $d = \sum_{i=1}^s i \dim U_i$ and $D = \sum_{i=1}^r i \dim V_i$.

Put $\mathcal{L}'_j = \oplus_{i=1}^j V_i$ and $K'_j = \oplus_{i=j}^s U_i$. By definition, we have $s \leq r$, $K'_j \subset \mathcal{L}_j$ and since

$$d = \sum_{i=1}^s i \dim U_i = \sum_{i=1}^s \dim K'_j,$$

and

$$D = \sum_{i=1}^r i \dim V_i = \sum_{i=1}^r \dim \mathcal{L}_i,$$

we indeed have $d \leq D$ and in fact $d = D$ if and only if G is stratified and X spans the first slice. Recall that d is always greater or equal than the topological dimension of G.

Furthermore, if $b \in \mathcal{B}$ and if we note

$$b_j = \mathrm{Vect}\{X_I \mid I \in b, |I| \leq j\}$$
$$b'_j = \mathrm{Vect}\{X_I \mid I \in b, |I| \geq j\},$$

then we deduce easily from the facts $b_j \subset K_j$, $b'_j \subset \mathcal{L}_j$, $d = \sum_{i=1}^s \dim K'_j$, $D = \sum_{i=1}^r \dim \mathcal{L}_j$ and $\omega(b) = \sum_{j=1}^r \dim b_j = \sum_{j=1}^r \dim b'_j$ that $d \leq \omega(b) \leq D$. In other words, one has $d = \inf\{\omega(b) \mid b \in \mathcal{B}\}$ and $D = \sup\{\omega(b) \mid b \in \mathcal{B}\}$.

IV.6 Sobolev's theorem

If $(G, \mathbf{X})$ is such that $d \leq D$, Proposition IV.4.1 and the volume estimates yield

$$\|T_t\|_{1 \to \infty} \leq CV(\sqrt{t})^{-1} \leq Ct^{-n/2}, \quad t > 0$$

if $d \leq n \leq D$. Together with Theorem II.2.7, this implies

IV.6.1 Theorem   *Let G be a nilpotent group and $\mathbf{X}$ a Hörmander system. Then:*

(i) *if $d \leq D$, for every $n \in [d, D]$, and every $p \in]1, +\infty[$, every $\alpha > 0$ such that $\alpha p < n$, $\Delta^{-\alpha/2}$ is bounded from L^p to $L^{pn/(n-\alpha p)}$. If $p = 1$, $\Delta^{-\alpha/2}$ is bounded from L^1 to $L^{n/(n-\alpha),\infty}$ and even, if $n \in]d, D[$, from L^1 to $L^{n/(n-\alpha)}$;*

(ii) *the same results still hold for every $n \geq d$ (or $n > d$) for the operators $(I + \Delta)^{-\alpha/2}$ and do so, whatever the position of D with respect to d.*

The boundedness of $\Delta^{-\alpha/2}$ from L^1 to $L^{n/(n-\alpha)}$ for $\alpha < n$, $n \in]d, D[$ follows from the weak result for $n = d$ and $n = D$ when $d < D$, thanks to the Marcinkiewicz theorem II.2.5.

Moreover, we have some kind of converse.

IV.6.2 Theorem   *Let G be a nilpotent group and $\mathbf{X}$ a Hörmander system. Let $1 < p < +\infty$, $0 < \alpha p < n$; then $\Delta^{-\alpha/2}$ is bounded from L^p to $L^{pn/(n-\alpha p)}$ if and only if $d \leq D$ and $n \in [d, D]$.*

Proof   It suffices to use Theorem II.3.1 to obtain, if $\Delta^{-\alpha/2}$ is bounded from L^p to $L^{pn/(n-\alpha p)}$, $\|H_t\|_{1 \to +\infty} = h_t(e) \leq Ct^{-n/2}$, and since $h_t(e) \geq V(\sqrt{t})^{-1}$, this implies that $d \leq D$ and $n \in [d, D]$.

IV.6.3 Remark   Recall that, if G is not simply connected, it may happen that $d > D$. In that case, one can show that, for $1 < p < +\infty$ and $0 < \alpha p < D$, $\Delta^{-\alpha/2}$ is bounded from L^p to $L^{pd/(d-\alpha p)} + L^{pD/(D-\alpha p)}$.

IV.7 Sobolev inequalities

In the previous section, we obtained in particular the inequality

$$\|\Delta^{-1/2}f\|_{2n/(n-2)} \leq C\|f\|_2, \quad \forall f \in L^2,$$

for a nilpotent group G, a Hörmander system $\mathbf{X}$ of local dimension $d > 2$, and of dimension at infinity D, and $n \in [d, D]$. Applying this inequality to $g = \Delta^{\frac{1}{2}}f$, $f \in C_0^\infty(G)$, and recalling that

$$\|\Delta^{\frac{1}{2}}f\|_2 = \left(\sum_{i=1}^{k} \|X_i f\|^2\right)^{\frac{1}{2}} \simeq \|\nabla f\|_2,$$

we see that for $n \in [d, D]$, $d > 2$,

$$\|f\|_{2n/(n-2)} \leq C\|\nabla f\|_2^2, \quad \forall f \in C_0^\infty(G). \tag{1}$$

We have seen in Chapter I (in $\mathbf{R}^n$, but the same argument works in G) that the inequality (1) is a consequence of

$$\|f\|_{n/(n-1)} \leq C\|\nabla f\|_1, \quad \forall f \in C_0^\infty(G). \tag{2}$$

In fact, (2) implies, for $p < n$,

$$\|f\|_{np/(n-p)} \le C\|\nabla f\|_p, \quad \forall f \in C_0^\infty(G).$$

In this section, we are going to deal with the strongest Sobolev inequality (2). We are first going to prove a very general result, which holds even outside of the nilpotent groups setting and which will apply here thanks to the estimates obtained in the previous sections.

IV.7.1 Theorem *Let G be a unimodular Lie group endowed with its Haar measure, $\mathbf{X} = \{X_1, ..., X_k\}$ a Hörmander system, h_t the heat kernel associated with $\mathbf{X}$, Δ the sublaplacian and ∇ the gradient associated with $\mathbf{X}$. Suppose there exists $n > 1$ and $C > 0$ such that*

$$\|h_t\|_\infty \le Ct^{-n/2}, \|X_i h_t\|_1 \le Ct^{-\frac{1}{2}}, \quad \forall t > 0, i \in \{1, ..., k\}. \tag{3}$$

Then

$$\|f\|_{n/(n-1)} \le C\|\nabla f\|_1, \forall f \in C_0^\infty(G).$$

If (3) is only satisfied for $0 < t \le 1$, then

$$\|f\|_{n/(n-1)} \le C\left(\|\nabla f\|_1 + \|f\|_1\right), \quad \forall f \in C_0^\infty(G).$$

Proof Recall that $H_t = e^{-t\Delta}$ and denote $\Delta_i = \Delta^{-1}X_i$, $i \in \{1, ..., k\}$. We have $\Delta^{-1} = \int_0^{+\infty} H_t \, dt$, and thus $\Delta_i f = \int_0^\infty H_t X_i f \, dt$, for all $f \in C_0^\infty(G)$. Notice that the dual operator of $H_t X_i$ is $-X_i H_t$ and thus that, by hypothesis,

$$\|H_t X_i\|_{1\to 1} = \|X_i H_t\|_{\infty\to\infty} = \|X_i h_t\|_1 \le Ct^{-\frac{1}{2}}, \quad \forall t > 0.$$

Moreover,

$$\|H_t X_i\|_{1\to+\infty} = \|X_i H_t\|_{1\to+\infty} = \|X_i h_t\|_\infty$$
$$\le \|X_i h_{t/2}\|_1 \|h_{t/2}\|_\infty \le Ct^{-(n+1)/2}, \quad \forall t > 0.$$

We can thus apply Remark II.2.8 to Δ_i. This gives

$$\xi(\{|\Delta_i f| > \lambda\}) \le C\left(\lambda^{-1}\|f\|_1\right)^{n/(n-1)}.$$

But, since $f = -\sum_{i=1}^k \Delta_i X_i f$, we get

$$\xi(\{|f| > \lambda\}) \le C[\lambda^{-1}\sum_{i=1}^k \|X_i f\|_1]^{n/(n-1)}, \quad \forall f \in C_0^\infty(G).$$

From now on, set $\|\nabla f\|_1 = \sum_{i=1}^k \|X_i f\|_1$. We have obtained a weak version of the result we are looking for, namely

$$\xi(\{|f| > \lambda\}) \le C(\lambda^{-1}\|\nabla f\|_1)^{n/(n-1)}, \quad \forall f \in C_0^\infty(G). \tag{4}$$

The next step of the proof consists in showing that, in this situation, the weak result (4) implies the strong result (2). Let us put

$$f_t = 1_{\{x \in G \mid f(x) > t\}},$$

for $f \geq 0$ in $C_0^\infty(G)$, and record the following two facts to be explained:

$$\|f\|_1 = \int_0^{+\infty} \|f_t\|_1 \, dt, \quad \|f\|_{n/(n-1)} \leq \int_0^{+\infty} \|f_t\|_{n/(n-1)} \, dt \qquad (5)$$

$$\|\nabla f\|_1 = \int_0^{+\infty} \|\nabla f_t\|_1 \, dt \qquad (6).$$

Indeed, we have (in the Bochner sense) $f = \int_0^{+\infty} f_t \, dt$ in the L^p spaces and (5) follows.

Moreover, we have, in the vector valued distribution sense,

$$\nabla f = \int_0^{+\infty} \nabla f_t \, dt,$$

where the integration actually takes place on a finite interval. By Sard's theorem, for almost every t, $\{x \in G \mid f(x) = t\}$ is either empty, or a hypersurface in G; ∇f_t is then a vector valued measure supported in this hypersurface, in particular the measures ∇f_t have disjoint supports. It follows that the total mass of the measure

$$\int_0^{+\infty} \nabla f_t \, dt = \nabla f,$$

i.e.

$$\|\nabla f\|_1 = \|\int_0^\infty \nabla f_t \, dt\|_1 = \int_0^{+\infty} \|\nabla f_t\|_1 \, dt,$$

which is nothing but (6).

Notice now that, by regularizing f and passing to the limit, we may apply to this function the weak inequality (4), if we interpret the L^1 norm of the gradient as the total mass of the measure ∇f_t. In particular,

$$\|f_t\|_{n/(n-1)} \leq [\xi(\{f_t > 1/2\})]^{(n-1)/n} \leq C\|\nabla f_t\|_1, \quad t > 0.$$

Using (5) and (6), one deduces from this inequality

$$\|f\|_{n/(n-1)} \leq \int_0^{+\infty} \|f_t\|_{n/(n-1)} \, dt \leq C \int_0^{+\infty} \|\nabla f_t\|_1 \, dt = \|\nabla f\|_1,$$

which is the announced inequality.

We now have to prove the second assertion of Theorem IV.7.1. Instead of Δ_i let us use $D_i = (I + \Delta)^{-1} X_i$, for which one obtains by the same method,

$$\xi(\{|D_i f| > \lambda\}) \leq C \left(\lambda^{-1} \|f\|_1\right)^{n/(n-1)}, \quad \forall f \in C_0^\infty(G).$$

Notice that $\sum_{i=1}^k D_i X_i = (I + \Delta)^{-1} \Delta = I - (I + \Delta)^{-1}$. Since, under the hypothesis $\|h_t\|_\infty \leq C t^{-n/2}$, $0 < t \leq 1$, the operator $(I + \Delta)^{-1}$ is bounded from L^1 to $L^{n/(n-2),\infty}$, and from L^1 to itself, it is also bounded from L^1 to $L^{n/(n-1)}$, $1 < n/(n-1) < n/(n-2)$, and we get

$$\xi(\{|f| > \lambda\}) \leq C\lambda^{-1} (\|f\|_1 + \|\nabla f\|_1)^{n/(n-1)}, \quad \forall f \in C_0^\infty(G).$$

Using the same procedure as before, we then obtain the strong inequality

$$\|f\|_{n/(n-1)} \leq C (\|f\|_1 + \|\nabla f\|_1), \quad \forall f \in C_0^\infty(G),$$

which ends the proof of Theorem IV.7.1.

Let us return to nilpotent groups. In this setting, the Harnack theorem IV.3.1, applied to $u(t, x) = h_t(x)$, tells us that

$$|X_i h_t(x)| \leq Ct^{-\frac{1}{2}} h_{2t}(x), \quad \forall t > 0, \forall x \in G,$$

which yields immediately $\|X_i h_t(x)\|_1 \leq Ct^{-\frac{1}{2}}$ for $t > 0$. On the other hand, for every $n \in [d, D]$ we have $\|h_t\|_\infty \leq Ct^{-n/2}$, using IV.4.1 and IV.5.6. We can thus apply Theorem IV.7.1 and obtain

IV.7.2 Theorem *Let G be a nilpotent group, $\mathbf{X}$ a Hörmander system, d the local dimension of $(G, \mathbf{X})$ and D the dimension at infinity of G. Then, for $n > 1$, we have:*

(i) $\|f\|_{n/(n-1)} \leq C\|\nabla f\|_1, \forall f \in C_0^\infty(G)$ *if and only if $d \leq D$ and $n \in [d, D]$.*

(ii) $\|f\|_{n/(n-1)} \leq C[\|\nabla f\|_1 + \|f\|_1], \forall f \in C_0^\infty(G)$ *if and only if $n \geq d$.*

The necessity of the conditions on n in (i) and (ii) may be seen by passing first to the corresponding estimates of $\|f\|_{2n/(n-2)}$ as at the beginning of this section, then by using Theorem IV.6.2. However we can now give a more elementary proof; it suffices to test inequality (i) on the function $\phi_t(x) = \sup\{0, t - \rho(x)\}$, suitably regularized, to get

$$t/2(V(t/2))^{(n-1)/n} \leq \|\phi_t\|_{n/(n-1)} \leq C\|\nabla \phi_t\|_1 \leq V(t), \quad t > 0,$$

therefore

$$t^n \leq CV(t)[V(t)/V(t/2)]^{n-1} \leq C'V(t), \quad t > 0.$$

By the estimates of the volume IV.5.8, this implies $d \leq D$ and $n \in [d, D]$. The same procedure gives the necessity of $n \geq d$ in (ii).

IV.7.3 Theorem *Let G be a nilpotent group, $\mathbf{X}$ a Hörmander system, d the local dimension of $(G, \mathbf{X})$, D the dimension at infinity of G, $p \in]1, +\infty[$ and $n < p$. Then we have:*

(i) *if $n \in [d, D]$, $|f(x) - f(xh)| \leq C\rho(h)^\alpha \|\nabla f\|_p$, where $\alpha = 1 - n/p$, for all $f \in C_0^\infty(G)$, $x, h \in G$;*

(ii) *if $n \geq d$, $|f(x) - f(xh)| \leq C\rho(h)^\alpha (\|\nabla f\|_p + \|f\|_p)$, where $\alpha = 1 - n/p$, for all $f \in C_0^\infty(G)$, $x, h \in G$.*

Proof Let us prove (i); (ii) follows along the same lines. Write

$$f(x) - f(xh) = f(x) - H_t f(x) - f(xh) + H_t f(xh) + H_t f(x) - H_t f(xh),$$

hence

$$|f(x) - f(xh)| \leq 2 \int_0^t \|\Delta H_s f\|_\infty \, ds + |H_t f(x) - H_t f(xh)|.$$

To estimate the first term, write $\Delta H_s f = \sum_{i=1}^k H_s X_i X_i f$. Now, IV.4.2 gives $\|H_s X_i\|_{p\to\infty} \leq s^{-1/2-n/2p}$, for $n \in [d, D]$. Therefore $\int_0^t \|\Delta H_s f\|_\infty \, ds \leq C t^{1/2-n/2p} \|\nabla f\|_p$. The second term is estimated by $\|H_t f(.) - H_t f(.h)\|_\infty \leq \rho(h)\|\nabla H_t f\|_\infty$. But, since $X_i H_t f = \sum_{j=1}^k \int_t^{+\infty} X_i H_s X_j X_j f \, ds$, one sees, using IV.4.2, that $\|\nabla H_t f\|_\infty \leq t^{-n/2p}\|\nabla f\|_p$. Finally

$$|f(x) - f(xh)| \leq 2C t^{1/2-n/2p}\|\nabla f\|_p + \rho(h)t^{-n/2p}\|\nabla f\|_p.$$

Choosing $t = \rho^2(h)$ yields (ii).

Let us finally introduce the following Sobolev seminorms:

$$\|f\|_{p,0} = \|f\|_p$$

and, for $\alpha \in \mathbf{N}^*$,

$$\|f\|_{p,\alpha} = \sum_{i=1}^k \|X_i f\|_{p,\alpha-1}, \, 1 \leq p \leq +\infty,$$

as well as the Hölder seminorms:

$$\Lambda_\alpha(f) = \sup_{h \in G} \frac{|f(x) - f(xh)|}{\rho(h)^\alpha}, \text{ if } 0 < \alpha < 1,$$

$$\Lambda_1(f) = \sup_{h \in G} \frac{|f(xh) - 2f(x) + f(xh^{-1})|}{\rho(h)},$$

$$\Lambda_\alpha(f) = \sup_{\substack{J \in \mathcal{I}(k) \\ |J|=\gamma}} \Lambda_{\alpha-\gamma}(X^J f), \text{ if } \gamma < \alpha \leq \gamma+1, \gamma \in \mathbf{N}^*.$$

We then have

IV.7.4 Corollary *Suppose that $d \leq D$ and $n \in [d, D]$. Let $1 \leq p < +\infty$ and $\alpha \in \mathbf{N}^*$. Then:*

(i) *if $\alpha p < n$, $\|f\|_{pn/(n-\alpha p)} \leq C\|f\|_{p,\alpha}$, $\forall f \in C_0^\infty(G)$;*

(ii) *if $\alpha p > n$, $\Lambda_{\alpha-n/p}(f) \leq C\|f\|_{p,\alpha}$, $\forall f \in C_0^\infty(G)$.*

IV.7.5 Corollary *Let $n \geq d$, $1 \leq p < +\infty$ and $\alpha \in \mathbf{N}^*$. Then:*

(i) *if $\alpha p < n$, $\|f\|_{pn/(n-\alpha p)} \leq C \sum_{\beta=0}^\alpha \|f\|_{p,\beta}$, $\forall f \in C_0^\infty(G)$;*

(ii) *if $\alpha p > n$, $\Lambda_{\alpha-n/p}(f) \leq C \sum_{\beta=0}^\alpha \|f\|_{p,\beta}$, $\forall f \in C_0^\infty(G)$.*

Proof Statement (i) in IV.7.4 with $\alpha = 1$ follows from IV.7.2 applied to f^s for f positive and $s = \frac{p(n-1)}{n-p}$ as in Section I.1. The case $\alpha \geq 2$ follows by iteration.

In order to prove IV.7.4 (ii), consider $J \in \{1, ..., k\}^\gamma$ with $\gamma < \alpha - n/p \leq \gamma + 1$. Set $g = X^J f$. One has $\|g\|_{p,\alpha-\gamma} \leq \|f\|_{p,\alpha}$ and $n/p < \alpha - \gamma \leq 1 + n/p$. If $n/p < 1$, then $\alpha - \gamma = 1$, and (ii) follows from IV.7.3 (i). If $n = p$, then $\alpha - \gamma = 2$. In this case, we have to show that

$$\Lambda_1(g) \leq C\|f\|_{n,2}.$$

This can be done by adapting the proof of IV.7.3 (i), taking into account the fact that the definition of $\Lambda_1(f)$ involves symmetric second differences.

Finally, if $\beta < n/p \leq \beta + 1$ for some $\beta \in \mathbb{N}^*$, we can use IV.7.4 (i) to see that $\|g\|_{q,\alpha-\beta-\gamma} \leq \|g\|_{p,\alpha-\gamma}$, where $n/q = n/p - \beta \leq 1$. Now the preceding cases yield $\Lambda_{\alpha-\gamma-n/p}(g) \leq C\|g\|_{p,\alpha-\beta-\gamma}$; since $\|g\|_{p,\alpha-\gamma} \leq \|f\|_{p,\alpha}$, this yields $\Lambda_{\alpha-\gamma-n/p}(X^J f) \leq C\|f\|_{p,\alpha}$. Taking the sup over J ends the proof of IV.7.4. The proof of IV.7.5 is similar.

References and comments

The main results presented in this chapter were first proved in [138], [142] and [144]. Observe however that there exists a vast literature, in the spirit of real analysis, on nilpotent groups that admit a dilation structure (see for example [49], [53] and the references given therein). For general facts on Lie groups (e.g. Section IV.1), a good reference is [126].

The Harnack estimates of Theorem IV.3.1 are the pivot of our analysis. The lifting procedure used to pass from stratified groups to general nilpotent ones is a purely algebraic trick which comes from [144]. Notice however that the lifting of fields to a free nilpotent Lie algebra had been used before by several authors, (e.g. [67]) and that a general lifting procedure for Hörmander fields had been elaborated in [104].

The volume estimates of Section IV.5 are due to Guivarc'h [59] for $t \geq 1$ and those for $0 < t \leq 1$ are contained in the results of [94]. Other references in that direction are [68], [144]. The presentation adopted here is taken from [108].

The Gaussian estimates on the heat kernel first appeared in [138], [144] and the presentation given here is from [152], [114]. The fundamental idea of the perturbation $e^{-\alpha\phi}e^{-t\Delta}e^{\alpha\phi}$ is due to Davies [40]. In [70], one finds also Gaussian estimates for related heat kernels. Interesting applications of those estimates in analysis are given in [105], [106], [23]. The best upper bound we know for the heat kernel on a nilpotent Lie group is

$$h_t(x) \leq CV(\sqrt{t}/(1 + \rho(x)/\sqrt{t})^{-1} \exp(-\rho^2(x)/4t);$$

(see [20], [43], [26]). The best lower bound has been obtained in [153], where it is proved that

$$h_t(x) \geq C_\varepsilon V(\sqrt{t})^{-1} \exp(-\rho^2(x)/(4 - \varepsilon)t),$$

for all $\varepsilon \in]0,1[$. As noticed in [63], one easily deduces from this tight two-sided gaussian bound that the Brownian motion X_t associated with Δ satisfies the law of the iterated logarithm

$$\limsup_{t \to +\infty} \frac{\rho(X_t)}{(2t \log \log t)^{1/2}} = 1 \quad a.s.$$

Extensions of the results of this chapter to nilmanifolds $M = G/H$, where G is a connected nilpotent Lie group and H a closed subgroup are studied in [84], [85].

The Sobolev inequalities obtained at the end of this chapter come from [138], [144]. For an alternative approach, in the setting of the Heisenberg group, see [97]. There are many further results concerning Sobolev inequalities on nilpotent Lie groups. In the case of Heisenberg groups, the best constant has been found by D. Jerison and J. Lee [69]. The best constant problem in the general case is open and interesting.

Other results and a simpler approach to Sobolev inequalities for stratified and homogeneous groups can be found in [107].

The Gagliardo-Nirenberg inequality for nilpotent Lie groups can be studied using the techniques of this chapter. In this context, the question is to characterize the multi-indices

$$(\omega(1), ..., \omega(k)) \in (\mathbb{R}^+)^k$$

for which we have

$$\|f\|_{n/(n-1)} \le C \prod_{i=1}^{k} \|X_i f\|_1^{\omega(i)/n}, \quad \forall f \in C_0^\infty(G)$$

with $n = \sum_{i=1}^{k} \omega(i)$. This has been done in [108].

CHAPTER V

LOCAL THEORY FOR
SUMS OF SQUARES OF VECTOR FIELDS

Let V be a C^∞ manifold of dimension N, $x \in V$, $\mathbf{X} = \{X_1, ..., X_k\}$ a Hörmander system of vector fields on V and ρ the distance on V associated to $\mathbf{X}$ as in Section III.4. We shall consider in this chapter the operator $\Delta = -\sum_{i=1}^{k} X_i^2$. Let B_t be the ball for ρ centred at x of radius t, and $V_x(t) = \xi(B_t)$, where ξ is a C^∞ non-vanishing measure on V. We shall give in Section 1 an estimate of $V_x(t)$, for x fixed and t small, by constructing a family of local dilations adapted to the situation, which will also yield in Section 3 the local Harnack inequalities uniform in t that we announced in Section III.2; in Section 5 we shall deduce from these inequalities some estimates of the heat kernel on V associated with Δ, for small time. More complete results, using the Harnack inequality and the volume estimates, will be presented in the case of unimodular Lie groups in Section 4.

V.1 Estimates of the volume

Let K_j, $j \in \mathbb{N}$, be the linear subspace of $T_x V$ spanned by the values at x of commutators of length at most j of vector fields of $\mathbf{X}$:

$$K_j = \text{Vect}\{X_I(x) \mid I \in \mathcal{I}(k), |I| \leq j\}.$$

The Hörmander condition implies the existence of $s \in \mathbb{N}$ such that

$$\{0\} = K_0 \subset K_1 \subset ... \subset K_s = T_x V.$$

Put $n_j = \dim K_j$, so that $0 = n_0 < n_1 < ... < n_s = N$, and $d = d_x = n_1 + 2(n_2 - n_1) + ... + s(n_s - n_{s-1}) = n_0' + ... + n_{s-1}'$, where $n_j' = N - n_j$ is the codimension of K_j; d is an integer greater than or equal to N. Notice that d_x is an upper semicontinuous function of x.

Our aim is to show

V.1.1 Theorem *There exists $C_x > 0$ such that $C_x^{-1} t^{d_x} \leq V_x(t) \leq C_x t^{d_x}$, for $0 < t < 1$.*

To this end, let us construct a basis of $T_x V$ adapted to the situation: choose a basis among the generators of K_1, complete it in a basis of K_2 using generators of K_2, and so on. The chosen vectors are thus of the form $Y_i(x)$, $i = 1, ..., N$, where Y_i is for $n_{j-1} < i \leq n_j$, a commutator of length j of vectors of $\mathbf{X}$. In particular, the vectors $Y_1, ..., Y_{n_1}$ belong to the family $\mathbf{X}$, but it may happen that $n_1 < k$. For y in the neighbourhood of x, $(Y_i(y))_{i=1}^{N}$ is a basis of $T_y V$, but not necessarily an adapted basis at y in the above sense.

The lower estimate of the volume is now within reach; indeed, let I_i be such that $Y_i = X_{I_i}$, $b = (I_1, ..., I_N)$, and ψ_b be the mapping built on b as in Section III.3. Let $R_t = \{\theta = (\theta_1, ..., \theta_N) \in \mathbb{R}^N \mid |\theta_i| \leq t^{|I_i|}, i = 1, ..., N\}$. The volume of R_t is t^d. Moreover, $\psi_b(R_t) \subset B_t$: the very way $\psi_b(\theta)$, $\theta \in R_t$, is written shows that it is possible to join it to x with a path of length smaller than t. Hence, since ψ_b is a local C^1-diffeomorphism, $\xi(B_t) \geq \xi(\psi_b(R_t)) \geq Ct^d$, for t small enough.

Let us now introduce a weight ω on $\{1, ..., N\}$ which will take into account the homogeneity of the fields Y_i with respect to the fields of $\mathbf{X}$: for $i \in \{1, ..., N\}$, put $\omega(i) = j$ if $n_j < i \leq n_{j+1}$; in other words, $\omega(i)$ is the length of the shortest expression of Y_i as a commutator of the X_j's. The integer $d = \sum_{i=1}^{N} \omega(i)$ then appears as the global homogeneity of an adapted basis in x with respect to the fields of $\mathbf{X}$.

To define the weight of a C^∞ function in the neighbourhood of x, put, for $I = (i_1, ..., i_p) \in \{1, ..., N\}^p$, $\omega(I) = \omega(I_1) + ... + \omega(I_p)$, and

$$\omega(f) = \inf\{m \mid \exists I_0 \in \mathcal{I}(N) \text{ satisfying } \omega(I_0) = m \text{ and } (Y^{I_0} f)(x) \neq 0\},$$

where $Y^I = Y_{i_1} \cdots Y_{i_p}$ if $I = (i_1, ..., i_p)$. If f and g are two C^∞ functions in the neighbourhood of x, we have $\omega(fg) = \omega(f) + \omega(g)$ and $\omega(f + g) \geq \inf(\omega(f), \omega(g))$.

Let $\mathbf{Y} = (Y_1, ..., Y_N)$. Since $\{Y_1(x), ..., Y_N(x)\}$ is a basis of $T_x V$, one can consider the exponential local coordinates built on $\mathbf{Y}$ at x. Denote those coordinates by $(u_1, ..., u_N)$, and by $\tilde{\partial}_i$ the C^∞ vector field in the neighbourhood of x given by $d\exp_{\mathbf{Y}, x}(\frac{\partial}{\partial u_i})$, $i = 1, ..., N$. If X is any C^∞ vector field in the neighbourhood of x, we can write in a unique way $X = \sum_{i=1}^{N} a_i \tilde{\partial}_i$, where the a_i's are C^∞ functions in the neighbourhood of x. We shall say that X is a good field at x if $\omega(a_i) \geq \omega(i) - 1$, $i = 1, ..., N$.

Let now $\{Z_1, ..., Z_N\}$ be a family of C^∞ vector fields in the neighbourhood of x. Let us write $Z_i = \sum_{j=1}^{N} a_{i,j} \tilde{\partial}_j$. We shall say that $\{Z_1, ..., Z_N\}$ is an admissible basis of vector fields at x if $a_{i,j}(x) = \delta_{i,j}$ and $\omega(a_{i,j}) \geq \omega(j) - \omega(i)$, $i, j = 1, ..., N$.

Let us assume for the moment the following

V.1.2 Key Lemma   *The basis Y is admissible at x; moreover, the fields $X_1, ..., X_k$ are good fields at x.*

The upper estimate of the volume follows from the second assertion of the key lemma. We shall use the first one later.

Let us consider the image Ω under $\exp_{\mathbf{Y}, x}$ of an open ball centred at 0; there exists $t_x > 0$ such that one can define, for $t \in [0, t_x[$, the map $\tilde{\phi}_t$ from Ω to itself by its expression in the chart $\phi_t(u_1, ..., u_N) = (t^{\omega(1)} u_1, ..., t^{\omega(N)} u_N)$.

If $t \in]0, t_x[$ and $i = 1, ..., k$, the field $X_i^t = td\tilde{\phi}_t^{-1}(X_i)$ is well defined on Ω:

$X_i^t(f)(x) = tX_i(f \circ \tilde{\phi}_t^{-1})(\tilde{\phi}_t(x))$, for $f \in C_0^\infty(\Omega)$; moreover,

$$X_i^t = t \sum_{j=1}^{N} a_{i,j}(\tilde{\phi}_t(x)) d\tilde{\phi}_t^{-1}\left(\tilde{\partial}_j\right)$$

$$= \sum_{j=1}^{N} a_{i,j}(\tilde{\phi}_t(x)) t^{1-\omega(j)} \tilde{\partial}_j, \quad i = 1, ..., k.$$

Now $\omega(a_{i,j}) \geq \omega(j) - 1$, since, by the key lemma, X_i is a good field. A simple application of Taylor's formula in the chart $\exp_{Y,x}^{-1}$ then shows that $a_{i,j}(\tilde{\phi}_t(x)) = O(t^{\omega(j)-1})$.

Finally, for $i = 1, ..., k$, the family $(X_i^t)_{t \in]0,t_x[}$ is bounded for the C^∞ topology on vector fields. It follows that the image under $\tilde{\phi}_t^{-1}$ of the ball of radius t for the distance induced by the system of vector fields $\mathbf{X} = \{X_1, ..., X_k\}$ — which is nothing but the ball of radius 1 for the distance induced by the system $\mathbf{X}^t = \{X_1^t, ..., X_k^t\}$ — is contained in a compact subset of V which is independent of t; we thus have

$$\xi\left(\tilde{\phi}_t^{-1}(B_t)\right) \leq C, \quad \forall t \in]0, t_x[.$$

Now it suffices to read $\tilde{\phi}_t^{-1}$ in the chart $\exp_{Y,x}^{-1}$ to see that its Jacobian equals Ct^{-d}. Finally, for $t \in]0, t_x[$, $t^{-d}V_x(t)$ is bounded above. Q.E.D.

V.2 Proof of the Key Lemma

Since $(Y_j)_{j=1,...,N}$ forms a basis of vector fields in the neighbourhood of x, there exist C^∞ functions $\alpha_{i,j}$ in the neighbourhood of x, such that

$$\tilde{\partial}_i = \sum_{j=1}^{N} \alpha_{i,j} Y_j, \quad i = 1, ..., N.$$

We are now going to use the asymptotic expansion of $\tilde{\partial}_i$ given in Section III.3.

The nth term of the expansion reads

$$(\text{ad}(u_i Y_1 + ... + u_N Y_N))^n Y_i = \sum_{\substack{I \in J(N) \\ |I|=n}} u_I (\text{ad } Y)_I Y_i$$

where $u_I = u_{i_1} \cdots u_{i_n}$ and

$$(\text{ad } Y)_I Z = [Y_{i_1}, [Y_{i_2}, ..., [Y_{i_n}, Z]...], \text{ if } I = (i_1, ..., i_n).$$

By Jacobi's identity, and the construction of the adapted basis at x, one can write $(\text{ad } Y)_I Y_i = \sum \lambda_A X_A$, where the λ_A's are constants and the X_A's are commutators of $X_1, ..., X_k$ of length $|A| = \omega(I) + \omega(i)$. Assume the following

V.2.1 Lemma   *If, for $I \in \mathcal{I}(k)$, we write $X_I = \sum_{j=1}^{N} f_j Y_j$, where the f_j's are C^∞ functions on Ω, then $\omega(f_j) \geq \omega(j) - |I|$.*

We have

$$X_A = \sum_{j=1}^{N} f_{j,A} Y_j,$$

with

$$\omega(f_{j,A}) \geq \omega(j) - |A| = \omega(j) - \omega(I) - \omega(i).$$

Now, for $i = 1, ..., N$, $\tilde{\partial}_i(y)$ may be written, for all $M \in \mathbb{N}^*$,

$$\sum_{j=1}^{M} \beta_{i,j}^M(y) Y_j(y) + O(|y|^{M+1})$$

with

$$\omega(\beta_{i,j}^M) \geq \omega(j) - \omega(i),$$

since $\omega(u_I) = \omega(I)$. It follows that $\omega(\alpha_{i,j}) \geq \omega(j) - \omega(i)$.

Let $M(y)$ be the matrix $(\alpha_{i,j}(y))_{1 \leq i,j \leq N}$. It satisfies $M(x) = \text{Id}$ and $\omega(\alpha_{i,j}) \geq \omega(j) - \omega(i)$.

Its inverse matrix, defined in the neighbourhood of x, has the same properties. As a matter of fact, let us call a matrix of C^∞ functions defined in the neighbourhood of x,

$$A(y) = (a_{i,j}(y))_{1 \leq i,j \leq N},$$

such that $\omega(a_{i,j}) \geq \omega(j) - \omega(i)$, *admissible*; the behaviour of the weight ω with respect to the sum and to the product of functions turns the set of admissible matrices into a closed algebra for the C^∞ topology on the coefficients.

Now $M(y)$ reads $\text{Id} + A(y)$, where A is admissible and $A(x) = 0$; hence, for y close to x, $M(y)^{-1} = \text{Id} - A(y) + A^2(y)...$ is admissible and, of course, $M(x)^{-1} = \text{Id}$. In other words, the first assertion of the key lemma is established, apart from Lemma V.2.1.

As to the X_i's, $i = 1, ..., k$, they read

$$X_i = \sum_{k=1}^{N} b_{i,k} Y_k,$$

with, using V.2.1, $\omega(b_{i,k}) \geq \omega(k) - 1$. Since

$$Y_k = \sum_{j=1}^{N} a_{k,j} \tilde{\partial}_j$$

with, as we just saw, $\omega(a_{k,j}) \geq \omega(j) - \omega(k)$, we indeed have

$$X_i = \sum_{j=1}^{N} c_{i,j} \tilde{\partial}_j,$$

with $\omega(c_{i,j}) \geq \omega(j) - 1$. This is the second assertion of the key lemma.

Let us now prove Lemma V.2.1. We have to show that, if $J \in \mathcal{I}(N)$,

$$\omega(J) < \omega(j) - |I| \;\Rightarrow\; (Y_J f_j)(x) = 0.$$

Let us consider the property

$H_n.$ $\qquad \omega(J) \leq n$ and $\omega(J) \leq \omega(j) - |I| \;\Rightarrow\; (Y_J f_j)(x) = 0.$

According to the construction of the adapted basis at x, for all $j \in \{1, ..., N\}$ such that $|I| < \omega(j)$, we have $f_j(x) = 0$; hence H_0 is true.

Let us now assume H_n, and let J_0 be such that $\omega(J_0) = n + 1$. Develop

$$(\operatorname{ad} Y)_{J_0} X = \sum_{j=1}^{N} (\operatorname{ad} Y)_{J_0}(f_j Y_j)$$

$$= \sum_{j=1}^{N} Y_{J_0}(f_j) Y_j + \sum_{j,K,L} Y_K(f_j)(\operatorname{ad} Y)_L Y_j.$$

In the second term, the summation ranges over $j \in \{1, ..., N\}$, K and L are the multi-indices with values in $\{1, ..., N\}$ such that $\omega(K) + \omega(L) \leq \omega(J_0) = n + 1$, and $\omega(K) \leq n$. Hence

$$(\operatorname{ad} Y)_{J_0} X(x) = \sum_{j=1}^{N} Y_{J_0}(f_j)(x) Y_j(x) + R$$

with, by the induction hypothesis,

$$R = \sum_{\omega(j) \leq \omega(K) + |I|} (Y_K f_j)(x)[(\operatorname{ad} Y)_L Y_j](x).$$

We have $R \in K_{n+|I|+1}$, since $(\operatorname{ad} Y)_L Y_j$ is a commutator of the X_i's of length

$$\omega(L) + \omega(j) \leq \omega(L) + \omega(K) + |I| \leq n + 1 + |I|.$$

In addition, $(\operatorname{ad} Y)_{J_0} X$ has length $\omega(I) + \omega(J_0) = |I_0| + n + 1$. Finally,

$$\sum_{j=1}^{N} Y_{J_0}(f_j)(x) Y_j(x) \in K_{n+|I|+1},$$

which amounts to saying that $Y_{J_0}(f_j)(x) = 0$ as soon as $\omega(j) \geq n + |I| + 1$. In other words, H_{n+1} holds. Lemma V.2.1 follows by induction.

V.3 Local scaling of the Harnack inequality

It follows from the Key Lemma V.1.2 that the system $\mathbf{X}^s = (X_i^s)_{i=1,...,k}$ satisfies the Hörmander condition on the neighbourhood Ω of $x \in V$, uniformly with respect to $s \in]0, t_x[$: assertion (1) of III.1.4 is a consequence of the fact that $X_1, ..., X_k$ are good fields, as explained just after the key lemma. As for

assertion (2), one considers the family of bases $\mathbf{Y}^s = (s^{\omega(j)} d\tilde{\phi}_s^{-1}(Y_j))_{j=1,\dots,N}$. It is easy to check that the fact that $\mathbf{Y}$ is an admissible basis yields (2)(ii) .
Denote by Δ^s the operator

$$-\sum_{i=1}^{k}(X_i^s)^2, \quad s \in \,]0,1].$$

Suppose for the sake of simplicity that $B_1 = B(x,1) \subset \Omega$ (otherwise, consider a smaller ball). Theorem III.2.4 tells us that, for $0 < c < 1$, t_1, t_2 such that $0 < t_1 < t_2 < +\infty$, $J \in \mathcal{I}(N)$ and $m \in \mathbb{N}$, $\exists C > 0$, $c \in \,]0,1[$ such that:

$$\forall s \in \,]0,t_x[, \forall \text{ positive } v \text{ satisfying } \left(\frac{\partial}{\partial t} + \Delta^s\right)v = 0$$

in $\mathbb{R}_+^* \times B_1$, we have

$$\sup_{y \in B_c} \left|\left(\frac{\partial}{\partial t}\right)^m \tilde{\partial}^J v(t_1, y)\right| \leq C \inf_{y \in B_c} v(t_2, y).$$

Here $\tilde{\partial}^J$ denotes the iterated field $\tilde{\partial}_{j_1}\dots\tilde{\partial}_{j_\ell}$, $J = (j_1, \dots, j_\ell)$. Of course, the constant C depends on the point x.

Now, if u is a positive solution of $\left(\frac{\partial}{\partial t} + \Delta\right)u = 0$ in $\mathbb{R}_+^* \times B_{\sqrt{s}}$, where $s \in \,]0,t_x[$, the function $v(t,y) = u(st, \tilde{\phi}_{\sqrt{s}}(y))$ is a positive solution of $(\frac{\partial}{\partial t} + \Delta^s)v = 0$ in $\mathbb{R}_+^* \times B_1$. On the other hand,

$$(\frac{\partial}{\partial t})^m v(t,y) = s^m (\frac{\partial}{\partial t})^m u(st, \tilde{\phi}_{\sqrt{s}}(y))$$

and

$$\tilde{\partial}^J v(t,y) = s^{\omega(J)/2} \tilde{\partial}^J u(st, \tilde{\phi}_{\sqrt{s}}(y)).$$

Therefore

$$\sup_{y \in B_{c\sqrt{s}}} \left|\left(\frac{\partial}{\partial t}\right)^m \tilde{\partial}^J u(st_1, y)\right| \leq C s^{-m-\omega(J)/2} \inf_{y \in B_{c\sqrt{s}}} u(st_2, y).$$

Finally, since $X_1, \dots, X_k$ are good fields, we see by induction that

$$X^J = \sum_I a_I \tilde{\partial}^I,$$

with $\omega(a_I) \geq \omega(I) - |I|$. We can thus state

V.3.1 Theorem   *Let V be a manifold, $\mathbf{X}$ a Hörmander system of vector fields on V, Δ the associated Laplacian, $x \in V$ and B_s the ball of radius s centred at x for the distance associated with $\mathbf{X}$. Then, for all $c \in \,]0,1[$, t_1, t_2*

such that $0 < t_1 < t_2 < +\infty$, $J \in \mathcal{I}(k)$ *and* $m \in \mathbb{N}$, *there exists* $C_x, t_x > 0$ *such that:*

$$\forall s \in]0, t_x[, \forall u \text{ a positive solution of } \left(\frac{\partial}{\partial t} + \Delta\right) u = 0,$$

in $\mathbb{R}_+^* \times B_{\sqrt{s}}$,

$$\sup_{y \in B_{c\sqrt{s}}} \left| \left(\frac{\partial}{\partial t}\right)^m X^J u(st_1, y) \right| \leq C_x s^{-m-|J|/2} \inf_{y \in B_{c\sqrt{s}}} u(st_2, y).$$

V.3.2 Remark Theorem V.3.1 extends verbatim to the case of the equation $\frac{\partial}{\partial t} + \Delta + Z = 0$, where $Z = \sum_{i,j=1}^{k} a_{i,j}[X_i, X_j] + \sum_{i=1}^{k} b_i X_i$, $a_{i,j}, b_i \in C_0^\infty(V)$.

V.4 The case of unimodular Lie groups

Suppose now that $V = G$ is a connected unimodular Lie group, endowed with a Haar measure, and that $\mathbf{X} = \{X_1, ..., X_k\}$ is a system of left invariant vector fields, satisfying the Hörmander condition at e, where e is the unit element of G, hence at every point.

Since the Carnot–Carathéodory distance is left invariant, a ball centred at x of radius t has the same volume as a ball of the same radius centred at e, and $V_x(t) = V_e(t) = V(t)$; by the way, the construction of an adapted basis gives the same result at every point: $d_x = d_e = d$.

The volume estimate then reads

V.4.1 Theorem *There exists $C > 0$ such that $\forall t \in]0, 1[$,*

$$C^{-1} t^d \leq V(t) \leq C t^d.$$

Moreover, if $x \in G$ and if $u(t, y)$ is a positive solution of $\left(\frac{\partial}{\partial t} + \Delta\right) u = 0$ in $I \times B(x, R)$ where I is some interval of $\mathbb{R}$ and where $R > 0$, the function $u(t, y) = u(t, x^{-1}y)$ is a positive solution of $\left(\frac{\partial}{\partial t} + \Delta\right) u = 0$ in $I \times B(e, R)$: it clearly follows that the constant which appears in V.3.1 is, in this setting, independent of x. In other words we have,

V.4.2 Theorem *Let G be a unimodular Lie group, $\mathbf{X} = \{X_1, ..., X_k\}$ a system of left invariant C^∞ vector fields on G, satisfying the Hörmander condition and Δ the associated Laplacian; for $x \in G$ and $s > 0$, let $B(x, s)$ be the ball centred at x of radius s for the distance associated with X. Then, $\forall c \in]0, 1[$, t_1, t_2 such that $0 < t_1 < t_2 < +\infty$, $J \in \mathcal{I}(k)$ and $m \in \mathbb{N}$, there exists $C > 0$ such that:*

$$\forall x \in G, \forall s \in]0, 1[, \forall u \text{ where } u \text{ is a positive solution of } \left(\frac{\partial}{\partial t} + \Delta\right) u = 0$$

in $\mathbb{R}_+^* \times B(x, \sqrt{s})$,

$$\sup_{y \in B(x,c\sqrt{s})} \left| \left(\frac{\partial}{\partial t} \right)^m X^J u(st_1, y) \right| \leq Cs^{-m-|J|/2} \inf_{y \in B(x,c\sqrt{s})} u(st_2, y).$$

We can now proceed as in Sections IV.4, IV.6 and IV.7 to obtain upper and lower estimates of the heat kernel, local Sobolev embedding theorems and local Sobolev inequalities.

V.4.3 Theorem *Let G, $\mathbf{X}$ and d be as above; let h_t be the associated heat kernel. Then there exists $c > 0$ such that, for all $t \in]0, 1[$, for all $x \in G$,*

$$\frac{1}{C} t^{-d/2} e^{-C\rho^2(x)/t} \leq h_t(x) \leq Ct^{-d/2} e^{-\rho^2(x)/Ct}.$$

V.4.4 Theorem *If Δ is the sublaplacian associated with $\mathbf{X}$, $1 \leq p < +\infty$, and $\alpha > 0$, such that $\alpha p < d$, then $(I + \Delta)^{-\alpha/2}$ is bounded from $L^p(G)$ to $L^q(G)$ if $p > 1$, $L^{q,\infty}(G)$ if $p = 1$, where q is defined by*

$$\frac{1}{q} = \frac{1}{p} - \frac{\alpha}{d}.$$

V.4.5 Theorem *If $p \in [1, d[$, there exists C_p such that*

$$\|f\|_{dp/(d-p)} \leq C_p \left(\|\nabla f\|_p + \|f\|_p \right), \quad \forall f \in C_0^\infty(G).$$

If $p \in]d, +\infty[$, there exists C_p such that

$$|f(x) - f(xh)| \leq C\rho(h)^{1-d/p} \left(\|\nabla f\|_p + \|f\|_p \right),$$

for all $f \in C_0^\infty(G)$, $x, h \in G$.

V.5 The general case

The results of the previous section seem to rely heavily on the group structure, which produces for free the necessary uniformity of the constants.

Nevertheless, we shall now show that, in the general setting of manifolds, the above theory, with a little additional work, yields in fact a scaled local Harnack estimate which is locally uniform with respect to the basis point x, hence a locally uniform central estimate of the heat kernel with respect to the volume $V_x(t)$.

The difficulty that one has to overcome is that the adapted basis may vary with the point x where it is constructed: in particular, d_x may suddenly jump down. However, a lifting procedure, due to Rothschild and Stein, (the proof has been simplified by Hörmander and Melin), will allow us to transfer this geometrically unstable situation into a stable one.

As above, let us consider a system of vector fields $\mathbf{X} = \{X_1, ..., X_k\}$ which satisfies the Hörmander condition on the manifold V. Suppose more precisely that these fields are s-Hörmander at $x \in V$, i.e., in the notation of Section V.1, that $K_s = T_x V$. Then there exists an integer m and vector fields $\widetilde{X}_1, ..., \widetilde{X}_k$ on $V \times \mathbb{R}^m$, such that

$$\widetilde{X}_i = X_i + \sum_{j=1}^{m} a_{i,j}(x, x') \frac{\partial}{\partial x'_j},$$

and that the $\widetilde{X}_i$'s are s-Hörmander at $(x, 0) \in V \times \mathbb{R}^m$, hence in a neighbourhood. Moreover, the $\widetilde{X}_i$'s are free of order s at $(x, 0)$. This means that the only linear relations between their commutators of length less than s, are those imposed by the Lie algebra structure. Of course, this remains true in a neighbourhood $\Omega_x \times \omega_x$ of $(x, 0) \in V \times \mathbb{R}^m$. In particular, all the linear relations between the commutators of length smaller than s of the $\widetilde{X}_i$'s remain the same in $\Omega_x \times \omega_x$: linear relations can only *disappear* suddenly, but the ones that take place are the compulsory ones! As a consequence, any adapted basis of $T_x V \times T_0 \mathbb{R}^m$ in the sense of Section 1, say

$$\widetilde{Y}_1(x, 0), ..., \widetilde{Y}_{N+m}(x, 0),$$

will propagate to $\Omega_x \times \omega_x$: thus $\widetilde{Y}_1(y, y'), ..., \widetilde{Y}_{N+m}(y, y')$ will be an adapted basis of $T_y V \times T_t \mathbb{R}^m$, for all $(y, y') \in \Omega_x \times \omega_x$.

We can now perform the local scaling of Section V.3 on the system

$$\tilde{\mathbf{X}}^{y, y'} = (\widetilde{X}_i(y, y'))_{i=1, ..., k},$$

uniformly in $(y, y') \in \Omega_x \times \omega_x$.

We obtain in this way a Harnack estimate similar to V.3.1 for the manifold $\Omega_x \times \mathbb{R}^m$, and the system $\tilde{\mathbf{X}} = (\widetilde{X}_1, ..., \widetilde{X}_k)$, for positive solutions of

$$\frac{\partial}{\partial t} - \sum_{i=1}^{k} \widetilde{X}_i^2,$$

in balls $\mathbb{R}_+^* \times B_{\sqrt{s}}(y, 0)$, $y \in \Omega_x$, $s < t_x$, with a constant depending only on x.

Now, this Harnack estimate transfers from $V \times \mathbb{R}^m$ to V: indeed, consider π, the canonical projection from $V \times \mathbb{R}^m$ onto V. Thanks to the splitting of $\widetilde{X}_i$ into X_i and a "vertical" field, we have $d\pi(\widetilde{X}_i) = X_i$, $i = 1, ..., k$. Thus the transformation $u \mapsto u \circ \pi$ lifts a positive solution of $\frac{\partial}{\partial t} - \sum_{i=1}^{k} X_i^2$ in a ball around y, $y \in \Omega$, into a positive solution of $\frac{\partial}{\partial t} - \sum_{i=1}^{k} \widetilde{X}_i^2$ in a ball around $(y, 0)$, that may be chosen of the same radius. We get in this way

V.5.1 Theorem *Let V be a manifold, $\mathbf{X}$ a system of Hörmander vector fields on V, Δ the associated Laplacian, and $x \in V$.*

Then, for all $c \in]0,1[$, t_1, t_2 such that $0 < t_1 < t_2 < +\infty$, $J \in \mathcal{I}(k)$ and $m \in \mathbf{N}$, there exists a neighbourhood Ω_x of x and $C_x, t_x > 0$ such that:

$$\forall s \in]0, t_x[, \forall y \in \Omega_x, \forall u, \text{ } u \text{ a positive solution of } \left(\frac{\partial}{\partial t} + \Delta \right) u = 0,$$

in $\mathbf{R}_+^* \times B_{\sqrt{s}}(y)$,

$$\sup_{z \in B_{c\sqrt{s}}(y)} \left| \left(\frac{\partial}{\partial t} \right)^m X^J u(st_1, z) \right| \leq C_x s^{-m-|J|/2} \inf_{y \in B_{c\sqrt{s}}} u(st_2, y).$$

The argument already used in IV.4 and V.4 now gives an estimate of the heat kernel. Indeed, let H_t be any semigroup such that $\left(\frac{\partial}{\partial t} + \Delta \right) H_t \varphi = 0$, $\forall \varphi \in C_0^\infty(V)$, $t > 0$, and let h_t be its kernel. We can state

V.5.2 Proposition   *For all $x \in V$, there exists a neighbourhood Ω_x of x and $C_x, t_x > 0$ such that*

$$h_t(y, z) \leq C_x V_z^{-1}(\sqrt{t}), \quad \forall t \in]0, t_x[, \forall y \in V, \forall z \in \Omega_x.$$

Notice that the variable y is free in the above estimate, because we can apply Harnack to the family of solutions $u_y(t, z) = h_t(y, z)$.

A simple application of the Heine-Borel lemma now shows that, if V is supposed to be a *compact* manifold, we have

V.5.3 Proposition   *There exists $C > 0$ such that*

$$\forall t \in]0,1[, \forall x, y \in V, \quad h_t(x, y) \leq C V_x^{-1}(\sqrt{t}).$$

The same conclusion holds if the system $\mathbf{X}$ satisfies the Hörmander condition uniformly on V. Finally, it is possible to get gaussian estimates for $h_t(x, y)$ as in V.4.3.

References and comments

This chapter stems entirely from [149]. However, many results had already been partially or totally proven, with other methods. For instance, the local volume estimate V.1.1 is contained in the work of Nagel, Stein and Wainger ([94]). In fact, their bound is more precise; in particular, they estimate $V_x(t)$ with respect to t and x.

The local Harnack estimate V.4.2 and the construction of the family of local dilations that is used in the proof come from [149]. A more general dilation theorem, which applies to subelliptic operators and not only to Hörmander sums of squares, has been obtained by Fefferman and Sanchez–Calle ([48]). Their result is harder and much deeper. It yields in particular estimates that are uniform with respect to the basis point. Together with the Harnack estimate which holds in this setting, (see References and Comments to Chapter III), it can be used as a basis for the study of these operators (see [152]). However, in the case of groups, where the uniformity with respect to the basis point is trivial, the method which is presented here suffices.

The small time uniform estimate V.5.2 of sums of squares heat kernels is due to Sanchez–Calle [116]. For a simple proof of the Rothschild–Stein lifting theorem, see [66].

For local gaussian estimates in the above context, see Jerison–Sanchez–Calle [70] and [71], Davies [41], and Varopoulos [152], [153].

CHAPTER VI

CONVOLUTION POWERS
ON FINITELY GENERATED GROUPS

VI.1 Introduction

This chapter and the next are not concerned with left invariant sublaplacians and their associated heat kernel h_t on unimodular Lie groups. Nevertheless, the matters we shall treat are closely related to the main stream of this book. Indeed, in the previous chapter we investigated the behaviour of $||h_t||_\infty$ for $0 \le t \le 1$. We would now like to study $||h_t||_\infty$ for $t \ge 1$. This will be achieved in Chapter VIII, but we are going to attack this problem from a somewhat more general point of view.

Let $F^{(k)}$ be the kth convolution power of $F \in L^1 \cap L^\infty$. In order to find out the behaviour of $||h_t||_\infty$ for $t \ge 1$, it suffices to look at $h_k = h_1^{(k)}$, $k = 1, 2, \dots$ Moreover, the function h_1 has a rapid decay at infinity since we know that $h_1(x) \le C \exp(-c\rho^2(x))$; see V.4.3. It is thus natural to address ourselves to the more general question of the behaviour of $||F^{(k)}||_\infty$, as k tends to infinity, for symmetric, positive compactly supported functions F of integral one.

Clearly enough, the Lie structure is no longer relevant here. Locally compact, unimodular groups which are compactly generated form the natural setting within which we will work. What we will eventually be able to show is that the decay of $||F^{(k)}||_\infty$ (with F as above) is governed by the volume growth of the group.

In this chapter we shall present some of the results which are central and for which we need the full thrust of our methods. We shall, for simplicity, concentrate on discrete finitely generated groups and not worry about minimal hypotheses. More general and complete results will be presented in the next chapter.

The next section contains the statements (without proofs) of known results concerning the volume growth of finitely generated groups (and more generally of compactly generated groups). These results are not used in an essential way in this book, but they are essential to see that the results we obtain give a rather complete picture of the situation. The main result is that finitely generated groups are either virtually nilpotent or of superpolynomial volume growth.

Sections 3 and 4 below are devoted to the study of convolution powers and Sobolev inequalities on finitely generated groups of superpolynomial growth. There, we introduce most of the main ideas and techniques used in our approach. The last section of this chapter contains the results concerning almost nilpotent finitely generated groups. An alternative proof of some of these results will be given in Chapter VII.

Let us insist on the rôle of the group structure in what follows. Let (X, ξ, ρ) be some nice measured metric space, for instance let X be a discrete, connected, countable graph, ξ be the counting measure, ρ the natural distance induced by the graph structure, and let us assume that each vertex in the graph has a finite bounded number of neighbours. Let $p^{(k)}(x, y) = \int_X p^{(k-1)}(x, z) \, dz$ be the kernel of the above Markov chain on $\ell^2(X)$. Also let $V(x, r)$ be the volume of the ball centred at x and of radius r. We might at first expect that an hypothesis like $V(x, r) \geq c r^A$ should imply some estimates of the form

$$p^{(k)}(x, y) \leq C k^{-B},$$

where $B = B(A)$ should be some number increasing with A. Indeed, in the case of the Cayley graph of a finitely generated group, we shall prove below that $B = A/2$. However, it is easy to show that such a result is simply not true in that generality. Before going back to our main subject, let us mention here that in the above general context we can show that, if X is not finite, i.e. if $V(x, r) \geq cr$, then $p^{(k)}(x, y) \leq C k^{-1/2}$. What can also be shown is that an isoperimetric inequality like

$$\sharp(\Omega) \leq C \left(\sharp(\partial \Omega) \right)^{A/(A-1)}$$

implies that $p^{(k)}(x, y) \leq C k^{-A/2}$; here Ω is any finite subset of X, $\sharp(\Omega)$ is the number of elements of Ω, $\partial\Omega$ is the boundary of Ω determined by the graph structure and the constant C does not depend on Ω.

VI.2 Distance and volume growth function on a group

Let Γ be a discrete, finitely generated group and let $\gamma_1, ..., \gamma_s$ be some fixed set of generators. For each $g \in \Gamma$ we can write $g = \gamma_{i_1}^{\varepsilon_1} \cdots \gamma_{i_m}^{\varepsilon_m}$ with $\varepsilon_k = \pm 1$, $i_k \in \{1, ..., s\}$ and some $n \in \mathbf{N}^*$. We can do this in many different ways and the number n of generators used to write g may change from one decomposition to another. Let $\rho(g)$ be the smallest possible n which occurs in all these decompositions of g. Of course, this definition of $\rho(g)$ depends on the choice of generators. But given a new set of generators, we obtain easily the essential invariance of $\rho(g)$:

$$C^{-1} \rho_{\text{new}}(g) \leq \rho_{\text{old}}(g) \leq C \rho_{\text{new}}(g), \quad g \in \Gamma,$$

for some $C > 0$ depending on the two sets of generators. (To see this, just write the new generators in terms of the old ones and vice versa.) The left invariant word distance on Γ is then defined by

$$\rho(x, y) = \rho(x^{-1}y), \quad x, y \in \Gamma,$$

and the growth function $n \mapsto V(n)$ of Γ by

$$V(n) = \sharp\{g \in \Gamma \mid \rho(g) \leq n\}.$$

The key theorem on the growth function is the following result, due to Gromov:

VI.2.1 Theorem   *If Γ is not virtually nilpotent, then for every $A > 0$ there exists $c = c_A > 0$ such that $V(n) \geq cn^A$, $n \geq 1$. In other words, the function V grows faster than every polynomial.*

We recall that a group Γ is nilpotent if there exists k_0 such that $\Gamma_{k_0} = \{e\}$ where

$$\Gamma = \Gamma_1 \supset \Gamma_2 \supset ... \supset \Gamma_k \supset ...$$

is the chain of commutator subgroups (i.e. $\Gamma_{k+1} = [\Gamma, \Gamma_k]$, $k \geq 1$). A group Γ is virtually nilpotent if it contains some nilpotent subgroup $\Gamma' \subset \Gamma$ such that $[\Gamma : \Gamma'] < +\infty$ (i.e. of finite index).

A fair amount of work has recently been done on the growth function of finitely generated groups. It is easy enough to find groups for which the function V grows like an exponential: all non-abelian free groups have that property.

What is highly non-trivial is to find groups of intermediate growth; that is a recent achievement of Grigorchuck. What Grigorchuck has proved is that there exists a finitely generated group and some $0 < \alpha < \beta < 1$, $c, C > 0$ such that:

$$ce^{n^\alpha} \leq V(n) \leq Ce^{n^\beta}, \quad n \geq 1.$$

Such a group is not easy to find; indeed previous results by Wolf, Tits and Guivarc'h show that solvable or linear groups are either of polynomial or of exponential growth.

For virtually nilpotent groups the growth function is very well understood and we have the following theorem of H. Bass (which was known before Gromov's theorem).

VI.2.2 Theorem   *Let Γ be a virtually nilpotent group. There exists $D = D(\Gamma) \in \mathbf{N}^*$ and $C > 0$ such that*

$$C^{-1}n^D \leq V(n) \leq Cn^D, \quad n \geq 1.$$

The interested reader could prove this result for himself as an exercise from the work of Chapter IV (where we investigated the volume growth of nilpotent Lie groups) and a classical theorem of Mal'cev which says that every discrete nilpotent group can be realized as a co-compact lattice in a nilpotent Lie group G (co-compact means that $\Gamma \subset G$ and G/Γ is a compact manifold). In fact, if Γ is virtually nilpotent with Γ' as a nilpotent subgroup of finite index and if

$$\Gamma' = \Gamma'_1 \supset \Gamma'_2 \supset ... \supset \Gamma'_k \supset ...$$

is the chain of commutator subgroups of Γ' as before, we have

$$D(\Gamma) = D(\Gamma') = \sum_1^\infty i[\Gamma'_k : \Gamma'_{k+1}].$$

Let us now remark that the above notions of distance and volume growth can easily be extended to the case of a locally compact group G as soon as it is compactly generated. Indeed, in this case, let $\Omega = \Omega^{-1}$ be a symmetric, compact neighbourhood of the identity element $e \in G$ such that Ω generates G (i.e. $G = \cup_1^\infty \Omega^n$). Let $\rho(g)$ be defined by

$$\rho(g) = \inf\{n \mid g \in \Omega^n\}, \quad g \in G,$$

and set

$$\rho(x,y) = \rho(x^{-1}y), \quad x,y \in G,$$
$$V(n) = \mathrm{Vol}(\Omega^n) = \mathrm{Vol}(\{g \in G \mid \rho(g) \le n\})$$

where $\mathrm{Vol}(A)$ is the Haar measure of $A \subset G$. In the case where G is a Lie group, the behaviour of V is described as follows.

VI.2.3 Theorem *For a Lie group G, there exists $C > 0$ such that either $C^{-1}n^D \le V(n) \le Cn^D$, $n \ge 1$, for some $D \in \mathbf{N}$, or $V(n) \ge C^{-1}e^{C^{-1}n}$, $n \ge 1$.*

This theorem is part of the work of Y. Guivarc'h, which also describes the algebraic structure of Lie groups of polynomial volume growth. It was also obtained around the same time by J. Jenkins. We emphasize here that Lie groups of intermediate growth in the sense of Grigorchuck (i.e. neither polynomial nor exponential) do not exist. Guivarc'h also showed that the above theorem holds if G is solvable instead of linear.

VI.3 The main results for superpolynomial groups

We shall say that Γ, a finitely generated group, is superpolynomial if $V(n) \ge cn^A$, $n \ge 1$, for all $A > 0$ and some $c = c_A > 0$. By Gromov's theorem, we know of course exactly which groups are not superpolynomial (i.e. the virtually nilpotent groups) and those groups will be considered separately in a later section. By the end, we will have achieved a rather complete understanding of the decay of convolution powers and Sobolev inequalities on finitely generated groups.

Let us introduce the Sobolev norm of a function $f \in C_0(\Gamma)$ (in the notation of Section VII.2):

$$S(f) = \sum_{\rho(x,y)=1} |f(x) - f(y)|$$

and its Dirichlet norm:

$$D(f) = \left(\sum_{\rho(x,y)=1} |f(x) - f(y)|^2 \right)^{\frac{1}{2}}.$$

In both summations above, we simply consider all pairs $(x, y) \in \Gamma^2$ such that $y = x\gamma$ for some $\gamma \in \{\gamma_1, ..., \gamma_s\}$. Both of these definitions depend on the choice of generators but, once more, by expressing old generators in terms of new ones, we see that changing the set of generators of Γ gives us an equivalent norm. We will have to consider also the Dirichlet norm associated with a symmetric probability measure ν on Γ and defined by

$$D_\nu(f) = \left(\frac{1}{2} \sum_{x,y} |f(x) - f(y)|^2 \nu(x^{-1}y) \right)^{\frac{1}{2}}.$$

In general there is no possible comparison between D_ν and D, but if the support of ν is finite and if ν changes some fixed set of generators, then we certainly have

$$C^{-1} D_\nu(f) \leq D(f) \leq C D_\nu(f), \quad f \in C_0(\Gamma).$$

The first main theorem on the subject is

VI.3.1 Theorem   *Let Γ be a superpolynomial finitely generated group. Then for all $N > 1$, there exists $C = C(N)$ such that*

$$\|f\|_{N/(N-1)} \leq C S(f), \quad f \in C_0(\Gamma).$$

In Gromov's terminology, the above theorem says that a superpolynomial Γ has infinite isoperimetric dimension. We can of course restate the above theorem in terms of the Dirichlet norm rather than the Sobolev norm.

VI.3.2 Theorem   *Let Γ be a superpolynomial group. Then for all $N > 2$, there exists $C = C(N)$ such that*

$$\|f\|_{2N/(N-2)} \leq C D(f), \quad f \in C_0(\Gamma).$$

This result could easily be deduced from the preceding one by using the elementary inequality $|x^\alpha - y^\alpha| \leq \alpha(x^{\alpha-1} + y^{\alpha-1})|x - y|$, valid for $\alpha \geq 1$ and $x, y > 0$, and Hölder inequalities.

In this chapter we are going to give a proof of Theorem VI.3.2 and leave Theorem VI.3.1 as an exercise for the reader. The other main theorem refers to convolution powers of measures on Γ and will also be proved here.

Let $\mu \in \mathbb{P}(\Gamma)$ be some probability measure on Γ which has the following properties:

(i) μ is symmetric, i.e. $\mu(g) = \mu(g^{-1})$, $g \in \Gamma$;

(ii) the support of μ is finite;

(iii) the support of μ generates Γ i.e. $\mathrm{Gp}(\mathrm{supp}(\mu)) = \Gamma$.

Of course, that last property means that there exist some fixed set of generators, $\gamma_1, ..., \gamma_s \in \Gamma$, such that $\mu(\gamma_j) > 0$, $j = 1, ..., s$.

VI.3.3 Theorem *Let Γ be a superpolynomial group and let $\mu \in \mathbb{P}(\Gamma)$ be a measure as above. Then the convolution powers $\mu^{(k)}$ of μ satisfy*

$$\sup_{x \in \Gamma} \mu^{(k)}(x) = O(k^{-A}) \quad \text{for all } A > 0.$$

The theorem says that, on a superpolynomial group, all the above measures have convolution powers which decrease superpolynomially. It is worth observing that $\mu^{(2k)}(e) = \sup_{x \in \Gamma} \mu^{(2k)}(x)$ and therefore what the above theorem really says is that $\mu^{(k)}(e) = O(k^{-A})$ for all $A > 0$.

The rest of this section will be devoted to the proof of a proposition which says, roughly speaking, that the conclusions of the last two theorems are equivalent properties.

VI.3.4 Proposition *Let ν be some symmetric probability measure on Γ and $A > 0$. The Dirichlet inequality*

$$\|f\|_{2A/(A-2)} \le C\, D_\nu(f), \quad f \in C_0(\Gamma)$$

is equivalent to the convolution powers decay

$$\nu^{(k)}(e) = O(k^{-A/2}).$$

Not only does this result cast some light on the links between Theorems VI.3.2 and VI.3.3, but it will be the main tool in the proof of those theorems. Note that in the above proposition ν need not be finitely supported and the support of ν does not need to generate Γ. Before proceeding with the proof, let us remark that we can always suppose that $\nu(e) > 0$. Indeed, if we set $\bar{\nu} = (\delta_e + \nu)/2$, where δ_e is the Dirac mass at the origin e of Γ, it is easy to see that $D_{\bar{\nu}} = \sqrt{2}^{-1} D_\nu$ and not too difficult to see that $\nu^{(k)}(e) = O(k^{-A/2})$ is equivalent to $\bar{\nu}^{(k)}(e) = O(k^{-A/2})$. This proves that we can always work with $\bar{\nu} = (\delta_e + \nu)/2$ instead of ν, or indeed suppose that $\nu(e) > 0$.

Let us prove that the convolution powers decay implies the Dirichlet inequality. An easy calculation (using the symmetry of the probability measure ν) shows that

$$D_\nu^2(f) = ((\delta_e - \nu) * f, f)$$

where $(.,.)$ is the scalar product on $\ell^2(\Gamma)$. From this it follows that D_ν^2 is the Dirichlet form associated with the symmetric submarkovian semigroup $T_t = e^{-t(\delta - \nu)}$ defined by

$$T_t f = e^{-t(\delta - \nu)} f = e^{-t} \sum_{0}^{\infty} \frac{(t\nu)^{(k)}}{k!} * f$$

on $\ell^2(\Gamma)$. But our hypothesis implies that

$$m_k = \sup_{x \in \Gamma}\{\nu^{(k)}(x)\} = O(k^{-A/2}).$$

On the other hand,

$$\|T_t\|_{1\to\infty} = \sup_x \{e^{-t} \sum_0^\infty \frac{(t\nu)^{(k)}(x)}{k!}\} \le e^{-t} \sum_0^\infty m_k t^k / k!$$

and the estimates on m_k imply

$$\|T_t\|_{1\to\infty} \le O(t^{-A/2}).$$

From this, our result is a direct application of the theory of Chapter II, and more particurlarly of Theorem II.5.2.

In order to prove that the Dirichlet inequality implies the convolution powers decay we will use the following elementary lemma, whose proof will be postponed until the end of this section:

VI.3.5 Lemma *Let $t_0 \ge t_1 \ge t_2... \ge t_k... > 0$ be a decreasing sequence of positive real numbers satisfying the difference inequality*

$$t_{k+1}^{1+2/N} \le C(t_k - t_{k+1}), \quad k \ge 0,$$

for some $N > 0$ and $C > 0$. Then there exists $C_1 > 0$, a numerical constant, such that

$$t_k \le (C_1 N C)^{N/2} k^{-N/2}, \quad k \ge 1.$$

In particular, $t_k = O(k^{-N/2})$ as k tends to infinity.

What we will need here is an apparently weaker form of Lemma VI.3.5 where the difference inequality is $t_k^{1+2/N} \le C(t_k - t_{k+1})$, and the conclusion is the same (since t_k tends to zero anyway it follows that $t_k^{1+2/N} - t_{k+1}^{1+2/N} = o(t_k - t_{k+1})$ and the two formulations are equivalent).

Let us suppose that the Dirichlet inequality $\|f\|_{2A/(A-2)} \le cD_\nu(f)$, $f \in C_0(\Gamma)$ holds. Let us fix $f \in C_0(\Gamma)$ such that $\|f\|_1 = 1$ and define the sequence

$$t_k = \|\nu^{(k)} * f\|_2^2 \quad k = 1, 2, ...$$

On the one hand, by our normalization $\|f\|_1 = 1$, we have also $\|\nu^{(k)} * f\|_1 \le 1$ and therefore, by the Hölder inequality,

$$\|\nu^{(k)} * f\|_{2A/(A-2)}^2 \ge \|\nu^{(k)} * f\|_2^{2(1+2/N)} = t_n^{1+2/N}.$$

On the other hand,

$$t_k - t_{k+1} = ((\delta - \nu^{(2)}) * \nu^{(k)} f, \nu^{(n)} * f) = D_{\nu^{(2)}}^2(\nu^{(k)} * f).$$

But, since we may assume that $\nu(e) > 0$, we have

$$\nu^{(2)}(x) \ge \nu(e)\nu(x), \quad x \in \Gamma.$$

Thus, using the preceding inequality and our Dirichlet inequality, we obtain

$$t_k - t_{k+1} = D^2_{\nu(2)}(\nu^{(k)} * f) \geq \nu(e) D^2_\nu(\nu^{(k)} * f) \geq c t_k^{1+2/N}.$$

Thanks to Lemma VI.3.5, this implies $t_k = O(k^{-A/2})$. After rescaling, we obtain $\|\nu^{(k)} * f\|_2 \leq Ck^{-A/4}\|f\|_1$, $f \in C_0(\Gamma)$, and by a usual duality argument,

$$\|\nu^{(k)} * f\|_\infty \leq Ck^{-A/4}\|f\|_2, \quad f \in C_0(\Gamma).$$

Finally, combining these two estimates, we get

$$\|\nu^{(k)} * f\|_\infty \leq Ck^{-A/2}\|f\|_1, \quad f \in C_0(\Gamma).$$

When $f = \delta_e$, this is the desired result.

Let us now give a proof of Lemma VI.3.5. Assume that the decreasing sequence of positive real numbers satisfies $t_{k+1}^{1+2/N} \leq C(t_k - t_{k+1})$, for some $N > 0$ and $C > 0$. Let k_0 be the first integer such that

$$\left(\frac{k+1}{k}\right)^{N/2} \leq 1 + C^{-1}t_0^{2/N}\left(\frac{k_0}{k+1}\right), \quad \text{for all } k \geq k_0.$$

It is easy to see that $k_0 = C_1 C N t_0^{-2/N}$ is enough, with C_1 a numerical constant. When $k \leq k_0$, we have $t_k \leq t_0 \leq t_0 k_0^{N/2} k^{-N/2}$. Suppose that $k \geq k_0$ is such that $t_k \leq C'k^{-N/2}$, and $t_{k+1} \geq C'(k+1)^{-N/2}$, where $C' = t_0 k_0^{N/2} = (C_1 C N)^{N/2}$. Then, the difference inequality satisfied by t_k and the definition of k_0 give us

$$t_{k+1} \leq t_k - C^{-1}t_{k+1}^{1+2/N}$$
$$\leq C'(k+1)^{-N/2}\left(\left(\frac{k+1}{k}\right)^{N/2} - C^{-1}t_0^{2/N}\left(\frac{k_0}{k+1}\right)\right)$$
$$\leq C'(k+1)^{-N/2}.$$

This shows that $t_k \leq (\frac{C_1 N C}{k})^{N/2}$ for all $k \geq 1$.

VI.4 Comparison of Dirichlet norms and finite variance

Let us pause and see what we have achieved up to now. We have seen that Dirichlet type inequalities are equivalent to convolution powers decay. We have also seen that Theorems VI.3.2 and VI.3.3 hold as soon as we can find, for each $A > 0$, some measure $\mu \in \mathbb{P}(\Gamma)$ which satisfies conditions (i), (ii) and (iii) of Section VI.3, and for which

$$\mu^{(k)}(e) = O(k^{-A}), \quad k \geq 1.$$

Attempts at constructing directly such measures have failed up to now. The reason for this is that, in the above, we required μ to be compactly supported (condition (i)). It is also clear that any non-trivial result on the decay of convolution powers of some compactly supported function has to depend on the fact that the group Γ is compactly (in fact finitely, here) generated.

The above programme, however, can be made to work if we consider non-compactly supported measures. The new notion that has to be introduced is that of the variance of a measure $\nu \in \mathbb{P}(\Gamma)$, denoted by

$$E(\nu) = \sum_{g \in \Gamma} \rho(g)^2 \nu(g).$$

Of course, $E(\nu)$ can easily be $+\infty$ for some $\nu \in \mathbb{P}(\Gamma)$ and $E(\nu) < +\infty$ if ν has finite support. But the class of measures of finite variance is larger than the class of measures of finite support, and we can prove by elementary calculations

VI.4.1 Proposition *Let Γ be some superpolynomial, finitely generated group. There exists $\nu \in \mathbb{P}(\Gamma)$, a symmetric probability measure which has finite variance (i.e. $E(\nu) < +\infty$) and for which the convolution powers decrease superpolynomially (i.e. such that $\nu^{(k)}(e) = O(k^{-A})$ for all $A > 0$).*

Let us suppose for a moment that the above probability measure ν has been constructed and that it also satisfies the Dirichlet norm comparison

$$D_\nu(f) \leq C D(f), \quad f \in C_0(\Gamma).$$

Then, applying VI.3.4 to the measure ν, we would get, for all $A > 2$,

$$\|f\|_{2A/(A-2)} \leq C_A D_\nu(f) \leq C'_A D(f), \quad f \in C_0(\Gamma),$$

which is the conclusion of Theorem VI.3.2.

The above Dirichlet norm comparison will be contained in the last proposition of this section where it will be proved that we indeed have

$$D_\nu(f) \leq C E(\nu) D(f), \quad f \in C_0(\Gamma),$$

for all symmetric $\nu \in \mathbb{P}(\Gamma)$.

Let us now return to the proof of VI.4.1. We really do the obvious thing and take ν to be a "radial" function which decays as slowly as possible. To be more precise, let χ_n be the characteristic function of the ball of radius $n \in \mathbb{N}^*$. We define ν by

$$\nu(g) = c \sum_1^\infty n^{-4} V(n)^{-1} \chi_n(g) = \sum \lambda_n V(n)^{-1} \chi_n(g)$$

where c is chosen so that ν has total mass one (i.e. $c^{-1} = \sum_1^\infty n^{-4}$) and $\lambda_n = cn^{-4}$. Observe that $E(\nu) \leq \sum n^2 \lambda_n < +\infty$. Now, using the basic fact that for any functions ϕ, ψ on Γ we have

$$\|\phi * \psi\|_\infty \leq \|\phi\|_1 \|\psi\|_\infty \quad \text{and} \quad \|\phi * \psi\|_\infty \leq \|\phi\|_\infty \|\psi\|_1,$$

some elementary combinatorics give

$$\nu^{(k)}(e) \le \left(\sum_{j=1}^{p-1} \lambda_j \right)^k + k \sum_{j \ge p} \lambda_j V(j)^{-1}.$$

Note that, for all $p \ge 1$,

$$\sum_{j=1}^{p-1} \lambda_j = 1 - \sum_{j \ge p} \lambda_j \le 1 - Cp^{-3}$$

and

$$\sum_{j \ge p} \lambda_j V(j)^{-1} \le CV(p)^{-1} p^{-3}.$$

From the above, we deduce at once

$$\nu^{(k)}(e) \le \left((1 - Cp^{-3})^k + CkV(p)^{-1} p^{-3} \right).$$

The final step consists in optimizing on p. Here it suffices to set $p = k^{1/4}$ to get

$$\nu^{(k)}(e) \le C \left[\exp(-ck^{1/4}) + k^{1/4} V(k^{1/4})^{-1} \right]$$

which shows that $\nu^{(k)}(e) = O(k^{-A})$ for every $A > 0$ as soon as $V(n) = O(n^{-A})$ for every $A > 0$.

We end this section with the statement and proof of the main result concerning comparisons of Dirichlet norms. As we indicated before, this result together with the preceding construction ends the proof of Theorems VI.3.2 and VI.3.3.

VI.4.2 Proposition *Let Γ be a finitely generated group and let μ be a probability measure on Γ; then there exists $C > 0$ which depends only on Γ and the set of fixed generators used to define ρ, for which*

$$D_\mu(f) \le C \, E(\mu) D(f), \quad f \in C_0(\Gamma).$$

The proof of this proposition is not difficult, but it is interesting because it introduces and uses the notion of "geodesic" (i.e. path of shortest length) in Γ. Let $x, y \in \Gamma$ be two fixed points. We shall say that the sequence $\gamma : x = x_0, x_1, ..., x_n = y$, is a path of length n which joins x to y if $\rho(x_j, x_{j+1}) = 1$ for all j. We shall say that such a path is a geodesic joining x to y if the length n of the path is equal to $\rho(x, y)$. Several such geodesics exist between x and y.

Let us fix x and y and let γ be a path between them, as above, though not necessarily geodesic. Also let $f \in C_0(\Gamma)$ be fixed. It is clear, from the Cauchy-Schwarz inequality, that

$$|f(x) - f(y)|^2 \le (\text{length of } \gamma) \sum |f(x_j) - f(x_{j+1})|^2.$$

Let us now, once and for all, fix for every $g \in \Gamma$ a geodesic $\gamma_g = (x_0, ..., x_n)$ joining e to g. The sequence $x\gamma_g = (xx_0, xx_1, ..., xx_g)$ is also a geodesic from x to xg. We therefore conclude that

$$|f(x) - f(xg)|^2 \le \rho(g) \sum_j |f(xx_j) - f(xx_{j+1})|^2.$$

Let us agree to denote

$$\nabla f(x) = \left(\sum |f(x) - f(x\gamma_j^\varepsilon)|^2 \right)^{\frac{1}{2}}$$

where the summation is over $\varepsilon = \pm 1$ and $\gamma_1, ..., \gamma_s$, the fixed set of generators. The preceding inequality gives then at once

$$|f(x) - f(xg)|^2 \le \rho(g) \sum_j |\nabla f(xx_j)|^2$$

It follows that

$$\sum_x |f(x) - f(xg)|^2 \le \rho(g) \sum_{x,j} |\nabla f(xx_j)|^2.$$

But, in the summation $\sum_{x,j}$, if we perform the summation over x first we obtain

$$\sum_{x,j} |\nabla f(xx_j)|^2 \le \rho(g) D^2(f)$$

so that

$$\sum_x |f(x) - f(xg)|^2 \le \rho(g)^2 D^2(f).$$

Our proposition follows upon observing that

$$D_\mu^2(f) = \frac{1}{2} \sum_{g,x} |f(x) - f(xg)|^2 \mu(g).$$

VI.5 Nilpotent finitely generated groups

In the preceding section, we were dealing with superpolynomial groups. By Gromov's theorem the picture will be completed if we look at virtually nilpotent ones. In what follows, D is the integer such that the growth functions of the virtually nilpotent group Γ satisfies (cf. Bass's theorem VI.2.2)

$$V(n) \simeq n^D, \quad n \ge 1.$$

VI.5.1 Theorem *Let Γ be a finitely generated virtually nilpotent group and $\mu \in \mathbb{P}(\Gamma)$ be a symmetric probability measure which satisfies conditions (i), (ii) and (iii) of Section VI.3. Then we have*

$$\sup_{x \in \Gamma} \mu^{(k)}(x) = O(k^{-D/2}).$$

Let us comment here that the result is sharp in the sense that there exists some $c > 0$ such that

$$\mu^{(2k)}(e) \geq ck^{-D/2}, \quad k \geq 1. \tag{1}$$

For the estimate (1), see References and Comments below.

Theorem VI.5.1 will be proved in the next chapter as a special case of a more general result. However, we are going to indicate here an alternative proof based on completely different ideas. Nevertheless, by Proposition VI.3.4 we know that it is equivalent to

VI.5.2 Theorem *Let Γ be a finitely generated virtually nilpotent group. We have, if $D > 2$,*

$$\|f\|_{2D/(D-2)} \leq C\, D(f), \quad f \in C_0(\Gamma).$$

It is not very difficult to obtain a proof of the two theorems above if we are prepared to use the results of Chapter IV, where we looked at nilpotent Lie groups, and Mal'cev's theorem (i.e., the fact that any nilpotent, finitely generated group is a co-compact lattice in some nilpotent Lie group). Indeed, the technique used to pass from a manifold to a co-compact lattice is presented in Chapter X, and can be applied here.

This approach also gives the *a priori* stronger

VI.5.3 Theorem *If Γ is as in Theorem VI.5.2, we have*

$$\|f\|_{D/(D-1)} \leq C\, S(f).$$

In particular, we obtain that Γ satisfies an isoperimetric inequality of dimension D: if A is a finite set in Γ and ∂A is the set of elements x of A for which there exists γ in the fixed set of generators such that $x\gamma \notin A$ or $x\gamma^{-1} \notin A$, then we have

$$\sharp(\partial A) \leq C(\sharp(A))^{(D-1)/D}.$$

As usual, it is possible to deduce the Sobolev inequality from the corresponding isoperimetric inequality and the two statements are equivalent.

VI.6 Kesten's conjecture

Let Γ be a finitely generated group and μ a symmetric probability measure on Γ with a finite and generating support. It is natural to ask whether the Markov chain given by the transition matrix $P(g, h) = \mu(g^{-1}h)$, $g, h \in \Gamma$ is transient or not. Equivalently, the problem is to find out whether the series $\sum_{k \geq 1} \mu^k(e)$ converges or not. This does not depend on the particular measure μ. It was H. Kesten who, for the first time, in the early 1960s, considered the above problem (hence the general name "Kesten's conjecture", which was vague but meant something to many people). The methods developed in this chapter together with Gromov's theorem give the answer immediately: as soon as we know the behaviour of $\mu^k(\{e\})$, we can decide whether the above series converges or not.

Theorem   *A finitely generated group Γ is recurrent (i.e. is not transient) if and only if $V(k) = O(k^2)$.*

Thus, by Gromov's theorem, such a group Γ admits a nilpotent finitely generated subgroup Γ' of finite index. Moreover, the volume growth exponent $D(\Gamma')$ must be at most 2. Furthermore, $D(\Gamma') = \sum_1^\infty k[\Gamma'_k : \Gamma'_{k+1}]$, where Γ'_{k+1} is the chain of commutator subgroups of Γ. This forces Γ to be Abelian. Finally, the only finitely generated recurrent groups are the finite extensions of $\{0\}$, $\mathbb{Z}$ and $\mathbb{Z}^2$. In the following chapter, we shall see that this result holds more generally for locally compact, compactly generated unimodular groups.

References and comments

The results presented here emerged first in a series of notes and papers of N. Varopoulos [127], [128], [129], [130], [131], [135], [137], [138], [141]; for Section 1, see [135]. It is worth emphasizing that this chapter contains, in a simple context, a great deal of the basic ideas used elsewhere in the book. As a matter of fact, the main ideas and techniques presented in this book were first developed around the material of this chapter.

The main references concerning the volume growth of groups are [89], [157], [13], [56], [57], [121], [122], [125], [59], [68], [54], [55]. The strong form of Gromov's theorem that we use is not stated explicitly in [56]. Nevertheless if one reads correctly Gromov's proof one obtains the present statement; see [141], [125], [88]. Bass' theorem appeared in [13]; see also [56], [122], [59]. See [54], [55] for Grigorchuck's example, and [86], [99] and [18] for Mal'cev's theorem.

There is a vast literature on random walks on groups (see [73], [74], and [61] where many further references can be found).

The results stated here concerning superpolynomial groups are, in some sense, far from optimal. For instance, if Γ is in fact of exponential growth (i.e. $V(n) \geq Ce^{cn}$), then for any measure μ satisfying conditions (i), (ii) and (iii) of Section VI.3, the estimate $\mu^k(e) = O(e^{-c'k^{1/3}})$ holds. We shall take up this theme in the next chapter.

It was pointed out to us by Y. Guivarc'h that, for nilpotent groups, the lower estimate (1) in Section VI.5 follows from our upper bound and a central limit theorem due to Raugi (see [100], [101]). The isoperimetric inequality VI.5.3 can also be obtained directly, as well as its superpolynomial analogue (see [37]).

Many partial results had been obtained in the direction of "Kesten's conjecture"; for example, it had been proved for connected Lie groups (see [60], [11] and [12]).

Observe that on proving the theorem in Section VI.6 (i.e. the solution of Kesten's conjecture) we only use Gromov's theorem for a very low growth: $O(k^2)$. It would be of interest to decide whether one could then directly prove that Γ has to be a finite extension of $\{0\}$, $\mathbb{Z}$ or $\mathbb{Z}^2$.

CHAPTER VII

CONVOLUTION POWERS
ON UNIMODULAR COMPACTLY GENERATED GROUPS

VII.1 Main results

We would now like to extend the results of Chapter VI to unimodular, locally compact, compactly generated groups. More precisely, we would like to relate the decay of $F^{(k)}(e)$, where F is some reasonable probability density on G, to the volume growth function V introduced in Section VI.2. Recall that V is defined by choosing a compact, symmetric, generating neighbourhood of $e \in G$, say Ω, and setting

$$V(n) = \mathrm{Vol}(\Omega^n) = \xi(\Omega^n), \quad n \geq 0,$$

where ξ is a fixed Haar measure on G. In the last chapter, we did this for finitely generated groups assuming that F is symmetric, compactly supported and charges a set of generators. Here we will prove

VII.1.1 Theorem *Let G be a unimodular compactly generated group. Let U be an open, generating neighbourhood of the origin e in G, and let F be a non-negative function such that $\int F = 1$, $\|F\|_\infty < +\infty$ and $\inf_U\{F\} \geq \varepsilon > 0$. Then, if $V(n) \geq cn^D$, we have*

$$\|F^{(k)}\|_\infty = O(k^{-D/2}),$$

and if $V(n) \geq ce^{cn^\alpha}$, we have

$$\|F^{(k)}\|_\infty = O(e^{-c'k^{\alpha/(\alpha+2)}}).$$

In fact, the techniques used to prove VII.1.1 are strong enough to give the following generalization:

VII.1.2 Theorem *Let G and U be as in VII.1.1. Let $F_1, ..., F_k, ...$ be a sequence of non-negative functions for which there exist $C, \varepsilon > 0$ such that $\int F_k = 1$, $\|F_k\|_\infty < C$, $\inf_U\{F_k\} \geq \varepsilon$. If $V(n) \geq cn^D$, we have*

$$\|F_1 * ... * F_k\|_\infty = O(k^{-D/2}).$$

If $V(n) \geq ce^{cn^\alpha}$, we have

$$\|F_1 * ... * F_k\|_\infty = O(e^{-c'k^{\alpha/\alpha+2}}).$$

In the next section we are going to develop a rather abstract machinery which is an adaptation of Chapter II to the discrete semigroup T^k, $k \in \mathbf{N}$, associated with a submarkovian operator T.

The third section contains comparison theorems for different Dirichlet forms.

Then, using all these results, we give a proof of Theorem VII.1.1 in the case where $V(n) \simeq n^D$.

The last section is devoted to a complete proof of Theorems VIII.1.1 and VIII.1.2, using methods which are adapted to more general volume growths.

VII.2 Dimension theory for symmetric submarkovian operators

Let (X, ξ) be a σ-finite measure space. We shall say that an operator T is *regularizing* if T is bounded on every L^p space, $1 \leq p \leq +\infty$, and bounded from L^1 to L^∞. A typical example of such an operator is the convolution by a bounded and integrable function on a locally compact group. Indeed, if $\phi \in L^1 \cap L^\infty$, one has

$$\|\phi * f\|_p \leq \|\phi\|_1 \|f\|_p, \, \forall f \in L^p(G), \quad 1 \leq p \leq +\infty;$$

and

$$\|\phi * f\|_\infty \leq \|\phi\|_\infty \|f\|_1, \, \forall f \in L^1(G).$$

Let T be a symmetric submarkovian operator on L^2; T is in particular a contraction of the spaces L^p, $1 \leq p \leq +\infty$. Let us denote by $P_t = P_t^T$ the symmetric submarkovian semigroup generated by $I - T$, i.e.

$$P_t = e^{-t} \sum_0^\infty (tT)^\nu / \nu!, \quad t > 0.$$

Paralleling Chapter II, we shall say that T satisfies (R_n) if it is regularizing and if $\|T^k\|_{1 \to \infty} \leq C k^{-n/2}$, $k \geq 1$. The elementary estimate

$$e^{-t} \sum_{\nu=1}^{+\infty} \frac{t^\nu \nu^{-n/2}}{\nu!} \leq C_n t^{-n/2}$$

shows that, for each regularizing operator S, there exists $C = C_S > 0$ such that

$$\|SP_t\|_{1 \to \infty} + \|P_t S\|_{1 \to \infty} \leq C t^{-n/2}, \quad t > 0 \tag{1}$$

as soon as T satisfies (R_n). In particular, if (X, ξ) is discrete, the identity operator is regularizing, and P_t satisfies (R_n) in the sense of Chapter II as soon as T satisfies (R_n) as above. Conversely, we can show that, if T is regularizing and

$$\|TP_t\|_{1 \to \infty} \leq C t^{-n/2}, \quad \forall t > 0,$$

then T satisfies (R_n) as above. We won't need this statement, so we leave its proof as an exercise to the reader.

Together with Remark II.2.8, inequality (1) gives

VII.2.1 Proposition *Let T be a symmetric submarkovian operator satisfying (R_n) and let S be some regularizing operator. Then, for every $p \in$*

$[1, +\infty[$, *with* α *such that* $0 < \alpha p < n$ *and* q *such that* $1/q = 1/p - \alpha/n$, *the operators* $S(I - T)^{-\alpha/2}$ *and* $(I - T)^{-\alpha/2}S$ *are bounded from* L^p *to* L^q *for* $p > 1$ *and from* L^1 *to* $L^{q,\infty}$.

VII.2.2 Remarks

(a) The operators $(I - T)^\alpha$, $\alpha \in \mathbb{R}$, can be defined in many equivalent ways. We can use semigroup theory, as in Chapter II, or spectral theory, or proceed in a more elementary way and set

$$(I - T)^\alpha = \sum_0^{+\infty} a_k(\alpha)T^k,$$

where $a_k(\alpha)$ is defined by

$$(1 - x)^\alpha = \sum_0^{+\infty} a_k(\alpha)x^k, \quad x \in [0, 1].$$

In particular, if T satisfies (R_n) with $n > 2$ and if S is a regularizing operator, we obtain

$$\|Sf\|_{2n/(n-2)} \leq C_S\|(I - T)^{\frac{1}{2}}f\|_2 = C_S((I - T)f, f)^{\frac{1}{2}},$$

which is a Dirichlet inequality involving the Dirichlet norm $((I - T)f, f)^{\frac{1}{2}}$.

(b) The presence in the left hand side of the above Dirichlet inequality of a regularizing operator S (for instance, in the group setting, convolution by a fixed function) is necessary. Indeed, since T is bounded on L^2, an inequality like

$$\|f\|_p \leq C\|(I - T)^{\frac{1}{2}}f\|_2, \, p > 2,$$

would imply $L^2(X, \xi) \subset L^p(X, \xi)$, which is impossible unless (X, ξ) is discrete. But precisely our aim in this chapter is also to consider non-discrete groups.

In order to obtain a converse to VII.2.1, we shall need a statement analogous to II.2.2, namely

VII.2.3 Proposition   *Let* T *be a regularizing symmetric submarkovian operator. The following statements are equivalent:*

(i) T *satisfies* (R_n);

(ii) *there exist* p, q *with* $1 \leq p < q \leq +\infty$ *and* C *such that*

$$\|T^k\|_{p \to q} \leq Ck^{-n(1/p-1/q)/2}, \quad \forall k \in \mathbb{N}^*.$$

Proof   Clearly (i) implies (ii) by interpolation. Suppose now that (ii) holds and put, for $f \in L^1 \cap L^q$ and $\ell \in \mathbb{N}^*$,

$$K(f, \ell) = \sup_{k \in \{1,\dots,\ell\}} \{k^{n/2q'}\|T^kf\|_q/\|f\|_1\}$$

where $1/q + 1/q' = 1$.

Let $\theta \in \,]0,1[$ be such that $1/p = \theta + (1-\theta)/q$, i.e. $1/p - 1/q = \theta/q'$. By hypothesis, for $k \in \mathbf{N}^*$,

$$\|T^{2k}f\|_q \le Ck^{-n\theta/2q'}\|T^k f\|_p.$$

On the other hand, Hölder's inequality and the boundedness of T on L^1 give

$$\|T^k f\|_p \le \|T^k f\|_1^\theta \|T^k f\|_q^{1-\theta} \le \|f\|_1^\theta \|T^k f\|_q^{1-\theta}.$$

Therefore

$$\|T^{2k}f\|_q \le Ck^{-n\theta/2q'}\|T^k f\|_q^{1-\theta}\|f\|_1^\theta$$
$$\le Ck^{-n/2q'} K(f,\ell)^{1-\theta}\|f\|_1$$

for all $k \in \{1,...,\ell\}$. Since T is regularizing, it follows that

$$\|T^k f\|_q \le Ck^{-n/2q'} K(f,\ell)^{1-\theta}\|f\|_1, \quad \forall k \in \{1,...,2\ell\},$$

whence

$$K(f,\ell) \le C\, K(f,\ell)^{1-\theta}$$

and

$$K(f,\ell) \le C^{1/\theta}.$$

Therefore $\|T^k\|_{1\to q} \le Ck^{-n/2q'}$, for all $k \in \mathbf{N}^*$. By duality, the same estimate holds for $\|T^k\|_{q'\to\infty}$, and, applying the same argument again, we obtain that

$$\|T^k\|_{1\to\infty} \le Ck^{-n/2}, \quad \forall k \in \mathbf{N}^*.$$

We now pass to the main result of this section, which is analogous to II.4.1.

VII.2.4 Theorem *Let T be a symmetric regularizing submarkovian operator on L^2. Then T satisfies (R_n) if and only if, for some (or every) $\alpha \in \,]0,n/2[$, $\|Tf\|_{2n/(n-2)} \le C\|(I-T)^{\alpha/2}f\|_2$, $f \in L^2$.*

Proof On account of VII.2.1, we only have to show that the above Dirichlet inequality implies (R_n).

Put $\tilde{T} = (I+T)/2$; $\tilde{T}$ is a submarkovian, symmetric operator on L^2 which is positive in the Hilbert space sense. Let $\tilde{T} = \int_0^1 \lambda\, dE_\lambda$ be its spectral decomposition. Since

$$\sup_{\lambda \in [0,1]} |\lambda^k - \lambda^{k+1}| \le 1/k,$$

it follows that $\|\tilde{T}^k - \tilde{T}^{k+1}\|_{2\to2} \le 1/k$, for all $k \in \mathbf{N}^*$. More generally, it can be shown in the same way that

$$\|(I-\tilde{T})^\beta \tilde{T}^k\|_{2\to2} \le C_\beta k^{-\beta}, \quad \forall k \in \mathbf{N}^*, \forall \beta > 0.$$

Moreover, $(I - T)^{\alpha/2} = c_\alpha (I - \tilde{T})^\alpha$, therefore the hypothesis implies

$$\|Tf\|_q \le C\|(I - \tilde{T})^{\alpha/2} f\|_2, \quad f \in L^2,$$

where $q = 2n/(n - 2\alpha)$. Thus, we have

$$\|T\tilde{T}^k f\|_q \le C\|(I - \tilde{T})^{\alpha/2} \tilde{T}^k f\|_2 \le Ck^{-\alpha/2}\|f\|_2,$$

and, since $\|(T\tilde{T})^k f\|_q \le \|T\tilde{T}^k f\|_q$,

$$\|T\tilde{T}\|_{2 \to q} \le Ck^{-\alpha/2}.$$

Therefore, according to Proposition VII.2.3, $T\tilde{T}$ satisfies (R_n). Now let $f \in L^1 \cap L^2$ be a positive function. Then, all the terms being positive, we have

$$\|(T\tilde{T})^k f\|_2 = ((T\tilde{T})^{2k} f, f) = \left(\frac{1}{2}\right)^{2k} \sum_0^{2k} C_{2k}^i (T^{i+2k} f, f)$$

$$\ge \left(\frac{1}{2}\right)^{2k} \sum_{j=0}^k C_{2k}^{2j} (T^{2j+2k} f, f).$$

But we also have

$$(T^{2j+2k} f, f) = \|T^{j+k} f\|_2^2 \ge \|T^{2k} f\|_2^2, \quad \text{for } j \le k,$$

therefore

$$\|T^{2k} f\|_2^2 \le C\|(T\tilde{T})^k f\|_2^2 \le Ck^{-n/2}\|f\|_1^2.$$

From this, we conclude by the usual duality argument (or again by VII.2.3) that T satisfies (R_n).

As a simple but typical application of the preceding results, we can obtain comparison results.

VII.2.5 Theorem   Let T_1, T_2 be two regularizing symmetric submarkovian operators. Suppose that T_1 satisfies (R_{n_1}) and that there exist $\alpha, \beta > 0$ such that

$$\|(I - T_1)^{\alpha/2} f\|_2 \le C\|(I - T_2)^{\beta/2} f\|_2, \quad f \in L^2.$$

Then T_2 satisfies (R_{n_2}) with $n_2 = \beta n_1/\alpha$.

Proof   We can suppose that $\alpha < n_1/2$. Indeed, the hypothesis

$$\|(I - T_1)^{\alpha/2} f\|_2 \le C\|(I - T_2)^{\beta/2} f\|_2, \quad f \in L^2,$$

implies, for every $0 < \delta < 1$,

$$\|(I - T_1)^{\alpha\delta/2} f\|_2 \le C\|(I - T_2)^{\beta\delta/2} f\|_2, \quad f \in L^2.$$

This can be proved either with the help of a classical lemma which states that if A and B are two positive operators on L^2,

$$A \leq B \Rightarrow A^\delta \leq B^\delta, \quad 0 < \delta < 1,$$

or by complex interpolation.

Now, since T_1 satisfies (R_{n_1}), we can use VII.2.1 to obtain the Dirichlet inequality

$$\|T_2 f\|_q \leq C\|(I - T_1)^{\alpha/2} f\|_2, \quad f \in L^2.$$

with $q = 2n_1/(n_1 - 2\alpha)$. The comparison hypothesis gives

$$\|T_2 f\|_q \leq C\|(I - T_2)^{\beta/2} f\|_2, \quad f \in L^2$$

and we conclude by Theorem VII.2.4.

Let us now give a simple result which is valid for operators that are neither necessarily submarkovian, nor symmetric and will be used in the proofs of Theorems VII.1.1 and VII.1.2.

VII.2.6 Lemma *Let T_k, $k = 1, \ldots$ be a sequence of operators which contract all the L^p spaces, $1 \leq p \leq +\infty$. Let us assume that these operators satisfy the family of inequalities:*

$$\|T_j f\|_2^{2+4/n} \leq C_1 \left(\|T_j f\|_2^2 - \|T_k T_j f\|_2^2 \right) \|f\|_1^{4/n}, \quad \forall f \in L^1 \cap L^2, \quad j, k \geq 1,$$

for some $n > 0$. Then we have

$$\|T_k \cdots T_1\|_{2 \to \infty} \leq \left(\frac{CC_1 n}{k} \right)^{n/4}, \quad \forall k \geq 1,$$

where C is a numerical constant. Moreover, if the adjoint operators T_k^* also satisfy the above hypotheses, we have

$$\|T_k \cdots T_1\|_{1 \to \infty} \leq \left(\frac{C' C_1 n}{k} \right)^{n/2}, \quad \forall k \geq 2,$$

where C' is a numerical constant.

Proof Fix a function f such that $\|f\|_1 = 1$ and set $t_k = \|T_k \cdots T_1 f\|_2^2$. The sequence t_k is non-increasing and satisfies by hypothesis

$$t_k^{1+2/n} \leq C_1(t_k - t_{k+1}), \quad k \geq 1.$$

It suffices to use Lemma VI.3.5 to conclude that $t_k \leq (\frac{CC_1 n}{k})^{n/2}$, $k \geq 1$. Let $C' = (CC_1 n)^{n/4}$.

After rescaling, we obtain

$$\|T_k \cdots T_1 f\|_2 \leq C' k^{-n/4} \|f\|_1,$$

or more generally

$$\|T_k \cdots T_{k'} f\|_2 \leq C' |k' - k|^{-n/4} \|f\|_1.$$

The same argument applied to the T_k^*'s yields by duality

$$\|T_k \cdots T_{k'} f\|_\infty \leq C' |k' - k|^{-n/4} \|f\|_2.$$

Putting together the above two estimates gives the conclusion.

VII.3 Comparison of Dirichlet forms

In this section, G will be a unimodular locally compact group, endowed with a Haar measure dx. We are going to obtain basic comparison results for Dirichlet forms. These results rely on a mean value inequality; see property $(\mathcal{H}_2)$ below and also Proposition VII.3.2. This inequality is essential to obtain a sharp relationship between the volume growth of a group and the decay of convolution powers. It is here that the group structure is going to play a fundamental rôle. It is here also that the assumption that G is compactly generated will be essential.

Let us introduce some notation. Let ρ be a given positive function on G; in the applications, $\rho(x)$ will be the distance from e to $x \in G$. Let Φ be another given positive function on G; in the applications, Φ will be the inverse of the volume of the smallest ball centred at e and containing x. Let us consider the following homogeneity property, which may or may not be satisfied by the pair (ρ, Φ):

$$\left. \begin{array}{c} \forall \gamma > 0, \exists C_\gamma \text{ such that }, \forall t > 0, \\ \displaystyle\int_{\rho(x) \leq t} \rho^\gamma(x) \Phi(x) \, dx \leq C_\gamma t^\gamma, \\ \text{and} \\ \displaystyle\int_{\rho(x) \geq t} \rho^{-\gamma}(x) \Phi(x) \, dx \leq C_\gamma t^{-\gamma}. \end{array} \right\} \qquad (\mathcal{H}_1)$$

Here is a typical example of a pair (ρ, Φ) on a group G which satisfies $(\mathcal{H}_1)$: let G be a stratified group — see Section IV.2 — (or even $G = \mathbb{R}^n$), $\rho(x) = |x|$ and $\Phi(x) = |x|^{-n}$, n being the homogeneous dimension of G.

Let $T_t = e^{-tA}$ be a symmetric submarkovian semigroup on $L^2(G, dx)$. Recall that T_t is a bounded analytic semigroup, hence that

$$\forall \alpha > 0, \quad \|A^\alpha T_t f\|_2 \leq C_\alpha t^{-\alpha} \|f\|_2, \quad \forall t > 0, \forall f \in L^2.$$

Let $\int_0^{+\infty} \lambda \, dE_\lambda$ be the spectral decomposition of A. For $f \in L^2$, we have

$$\|AT_t f\|_2^2 = \int_0^{+\infty} \lambda^2 e^{-2t\lambda} \, d(E_\lambda f, f)$$

and

$$(A^\alpha f, f) = \|A^{\alpha/2} f\|_2^2 = \int_0^{+\infty} \lambda^\alpha d(E_\lambda f, f).$$

It follows immediately that, for $0 < \alpha < 2$

$$\|A^{\alpha/2}f\|_2^2 = c_\alpha \int_0^{+\infty} \left(t^{1-\alpha/2}\|AT_tf\|_2\right)^2 \frac{dt}{t}.$$

Given A and ρ as above, we introduce the following property, which may or may not be satisfied by the pair (A, ρ):

$$\int_G |f(xh) - f(x)|^2 \, dx \le C\rho^2(h)\|A^{1/2}f\|_2^2, \quad f \in \mathcal{D}(A^{1/2}). \qquad (\mathcal{H}_2)$$

Property $(\mathcal{H}_2)$ is in fact an L^2 version of the mean value inequality. A typical example of a pair (A, ρ) satisfying $(\mathcal{H}_2)$ is obtained by considering a unimodular Lie group G, a Hörmander system $\mathbf{X} = \{X_1, ..., X_k\}$ of left invariant vector fields, $A = -\sum_{i=1}^k X_i^2$, and ρ the distance associated with $\mathbf{X}$ as in Section III.4. For a proof, see Proposition VIII.1.1 below.

The following result shows how properties $(\mathcal{H}_1)$ and $(\mathcal{H}_2)$ yield comparisons of Dirichlet forms.

VII.3.1 Proposition *Let G, ρ, Φ, $T_t = e^{-tA}$ be as above. Suppose that (ρ, Φ) satisfies $(\mathcal{H}_1)$ and (A, ρ) satisfies $(\mathcal{H}_2)$. Then, for every $0 < \alpha < 1$, we have*

$$\int \left(|f(xh) - f(x)|/\rho^\alpha(h)\right)^2 \Phi(h) \, dh \, dx \le C_\alpha \|A^{\alpha/2}f\|_2^2, \quad \forall f \in \mathcal{D}(A^{\alpha/2}).$$

Proof Let $f_h(x) = f(xh)$, for $x, h \in G$ and write

$$f_h - f = \int_0^t \left((AT_sf)_h - AT_sf\right) ds + (T_tf)_h - T_tf.$$

Taking the L^2 norm, we obtain

$$\|f_h - f\|_2 \le 2\int_0^t \|AT_sf\|_2 \, dx + \|(T_tf)_h - T_tf\|_2.$$

According to $(\mathcal{H}_2)$, we have

$$\|(T_tf)_h - T_tf\|_2 \le C\rho(h)\|A^{1/2}T_tf\|_2 = C\rho(h)\|\int_t^{+\infty} A^{3/2}T_sf \, ds\|_2$$

$$\le C'\rho(h) \int_{t/2}^{+\infty} s^{-1/2}\|AT_sf\|_2 \, ds.$$

The last inequality follows from the fact that, since T_t is a bounded analytic semigroup,

$$\|A^{3/2}T_sf\|_2 \le Cs^{-1/2}\|AT_{s/2}f\|_2.$$

From the above, we deduce that

$$\|f_h - f\|_2 \le 2\int_s^t \|AT_sf\|_2 \, ds + c'\rho(h) \int_{t/2}^{+\infty} s^{-1/2}\|AT_sf\|_2 \, ds.$$

Choose $t = \rho^2(h)$ and set

$$K(h,s) = C\left(s^{\alpha/2}\rho^{-\alpha}(h)1_{\{s \le \rho^2(h)\}} + s^{-(1-\alpha)/2}\rho^{1-\alpha}(h)1_{\{2s \ge \rho^2(h)\}}\right)$$

and

$$g(h) = \int_0^{+\infty} K(h,s)\left(s^{1-\alpha/2}\|AT_s f\|_2\right) \frac{ds}{s}.$$

We have

$$\int_G (\|f_h - f\|_2/\rho^\alpha(h))^2 \, \Phi(h) \, dh \le \int_G g^2(h)\Phi(h) \, dh.$$

Now, g is of the form

$$g(h) = T_K\phi(h) = \int_0^{+\infty} K(h,s)\phi(s) \frac{ds}{s},$$

where $\phi(s) = s^{1-\alpha/2}\|AT_s f\|_2$. We claim that T_K is a bounded operator from $L^2(\mathbb{R}^{+*}, s^{-1}\,ds)$ to $L^2(G, \Phi(h)\,dh)$. From this claim and the above it follows that

$$\int (\|f_h - f\|_2/\rho^\alpha(h))^2 \, \Phi(h) \, dh \le C \int_0^{+\infty} \phi(s)^2 \frac{ds}{s}$$

$$= C \int_0^{+\infty} \left(s^{1-\alpha/2}\|AT_s f\|_2\right)^2 \frac{ds}{s}$$

which is the conclusion of VII.3.1, since we have seen that

$$\int_0^{+\infty} \left(s^{1-\alpha/2}\|AT_s f\|_2\right)^2 \frac{ds}{s} = c_\alpha^{-1}\|A^{\alpha/2}f\|_2^2.$$

We are therefore left with the claim to prove. We have

$$\int_0^{+\infty} K(h,s) \frac{ds}{s} \le C_1$$

and, using $(\mathcal{H}_1)$,

$$\int_G K(h,s)\Phi(h) \, dh \le C_2,$$

where C_1 is independent of $h \in G$ and C_2 independent of $s \in \mathbb{R}^{+*}$. The boundedness of T_K from $L^2(\mathbb{R}^{+*}, s^{-1}\,ds)$ to $L^2(G, \Phi(h)\,dh)$ follows easily from these estimates.

We are now going to see that Proposition VII.3.1 can be applied in the study of convolution powers on compactly generated groups. Let G be a locally compact, unimodular, compactly generated group. Let Ω be a fixed neighbourhood of $e \in G$ that is symmetric, compact and generating. Let ρ be defined as in Section VI.2 by

$$\rho(h) = \inf\{n \in \mathbb{N} \mid h \in \Omega^n\}$$

and
$$V(n) = \text{Vol}(\Omega^n).$$

Fix the following notation for the remainder of this chapter: $F_0 = |\Omega^3|^{-1}1_{\Omega^3}$, T_0 is the operator of right convolution by F_0, and $A = I - T_0$.

More generally, let F be a non-negative, bounded function on G such that $\int F(x)\, dx = 1$, and $F(x) = F(x^{-1})$, for all $x \in G$. Let T be the operator associated with the right convolution by F, i.e.

$$Tf(x) = \int F(y^{-1}x)f(y)\, dy.$$

Clearly, T is symmetric, markovian, regularizing and left invariant. Moreover, $\|(I - T)^{1/2}f\|_2^2$ can be written

$$
\begin{aligned}
((I - T)f, f) &= \int f^2(x)\, dx - \int\int F(y^{-1}x)f(y)f(x)\, dy\, dx \\
&= \frac{1}{2}\int\int F(y^{-1}x)\,(f(x) - f(y))^2\, dy\, dx \\
&= \frac{1}{2}\int\int (f(xh) - f(x))^2\, F(h)\, dh\, dx.
\end{aligned}
$$

The next proposition shows that (A, ρ) satisfies property $(\mathcal{H}_2)$.

VII.3.2 Proposition G, Ω, ρ *being as above, we have, for all $f \in C_0(G)$ and all $h \in G$,*

$$
\begin{aligned}
\|f_h - f\|_2^2 &\leq |\Omega|^{-1}\rho^2(h)\int |f(xh) - f(x)|^2 1_{\Omega^3}(y)\, dy\, dx \\
&= C_\Omega \rho^2(h)\|(I - T_0)^{1/2}f\|_2^2.
\end{aligned}
$$

Proof Recall that $f_h(x) = f(xh)$. Let $h \in G$ be fixed and set $\rho(h) = n$. We can find $e = h_0, h_1, ..., h_n = h$ such that $h_{j+1} \in h_j\Omega$, $j = 0, ..., n - 1$. Moreover, for any $x_1, ..., x_{n-1}$ we have

$$|f(h) - f(e)|^2 \leq n\left(|f(x_1) - f(e)|^2 + ... + |f(x_{n-1} - f(h)|^2\right).$$

Integrating against the measure $dx_1 \cdots dx_{n-1}$, over the set $h_1\Omega \times h_2\Omega \times ... \times h_{n-1}\Omega$, we obtain

$$
\begin{aligned}
|f(h) - f(e)|^2 \leq{}& n(|\Omega|^{-1}(\int_{h_1\Omega} |f(x_1) - f(e)|^2\, dx_1 \\
&+ |\Omega|^{-1}\int_{h_1\Omega}\int_{h_2\Omega} |f(x_2) - f(x_1)|^2\, dx_1\, dx_2 + \\
&+ ... + \int_{h_{n-1}\Omega} |f(x_{n-1}) - f(h)|^2\, dx_{n-1})).
\end{aligned}
$$

Note that

$$
\begin{aligned}
\int_{h_1\Omega} |f(x_1) - f(e)|^2\, dx_1 &\leq \int_{\Omega^3} |f(y) - f(e)|^2\, dy \\
&= \int_G\int_G |f(zy) - f(z)|^2 1_{\Omega^3}(y)\, dy\, d\mu_0(z)
\end{aligned}
$$

where $\mu_0 = \delta_e$ is the Dirac mass at the origin $e \in G$. In the same way,

$$\int_{h_{n-1}\Omega} |f(x_{n-1}) - f(h)|^2 \, dx_{n-1} \leq \int_{\Omega^3} |f(hy) - f(h)|^2 \, dy$$
$$= \int_G \int_G |f(zy) - f(z)|^2 1_{\Omega^3}(y) \, dy \, d\mu_{n-1}(z)$$

where $\mu_{n-1} = \delta_h$, and

$$|\Omega|^{-1} \int_{h_i\Omega} \int_{h_{i+1}\Omega} |f(x_{i+1}) - f(x_i)|^2 \, dx_i \, dx_{i+1}$$
$$\leq |\Omega|^{-1} \int_{y_i\Omega} \int_{\Omega^3} |f(x_iy) - f(x_i)|^2 \, dx_i \, dy$$
$$= \int_G \int_G |f(zy) - f(z)|^2 1_{\Omega^3}(y) \, dy \, d\mu_i(z)$$

where $d\mu_i(z) = |\Omega|^{-1} 1_{\Omega^3}(z) \, dz$. Finally, we have

$$|f(h) - f(e)|^2 \leq n \int_G \int_G |f(zy) - f(z)|^2 1_{\Omega^3}(y) \, dy \, d\mu_h(z)$$

where $\mu_h = \sum_{i=0}^{n-1} \mu_i$ satisfies $\|\mu_h\| = n$. Here $\|\mu_h\|$ is the total mass of the measure μ_h. Applying the above to the function $z \mapsto f(xz)$ and integrating over G with respect to x against the Haar measure dx leads to

$$\|f_h - f\|_2^2 \leq |\Omega|^{-1} n \|\mu_h\| \int_G \int_G |f(xy) - f(x)|^2 1_{\Omega^3}(y) \, dy \, dx$$
$$= 2|\Omega^3| |\Omega|^{-1} \rho^2(h) \|(I - T_0)^{1/2} f\|_2^2,$$

where we used the right invariance of the Haar measure dx to obtain the inequality.

As far as property $(\mathcal{H}_1)$ is concerned, we have

VII.3.3 Proposition *Assume that $V(n) \simeq n^D$. Let $\Phi(x) = \rho(x)^{-D}$, $x \in G$. Then (ρ, Φ) satisfies $(\mathcal{H}_1)$, that is,*

$$\int_{\rho \leq t} \rho^\gamma \Phi \leq Ct^\gamma, \qquad \int_{\rho \geq t} \rho^{-\gamma} \Phi \leq Ct^{-\gamma}$$

for all $\gamma > 0$, $t > 0$.

Proof We shall show the first estimate and leave the other as an exercise. Note first that for $t < 1$, $\int_{\rho \leq t} \rho^\gamma \Phi = 0$! For $t \geq 1$, let N be the smallest integer such that $2^N \geq t$. We have

$$\int_{\rho \leq t} \rho^\gamma \Phi \leq \int_{\rho \leq 1} \rho^\gamma \Phi + \sum_{i=1}^N \int_{2^{i-1} < \rho < 2^i} \rho^\gamma \Phi$$
$$\leq C \sum_{i=0}^N 2^{i\gamma} \leq C' 2^{N\gamma} \leq C'' t^\gamma.$$

Finally, Propositions VII.3.2, VII.3.3 and VII.3.1 yield

VII.3.4 Proposition *Let G be a unimodular compactly generated locally compact group. Suppose that its volume growth function V is such that $V(n) \simeq n^D$. Let Ω, ρ and T_0 be as above. Then, for every $0 < \alpha < 1$, we have*

$$\int |f(xh) - f(x)|^2 \rho^{-2\alpha-D}(h)\, dh\, dx \leq C_\alpha \|(I - T_0)^{\alpha/2} f\|_2^2, \quad \forall f \in \mathcal{D}(A^{1/2}).$$

Some comments on this result are in order. Let F_α be any probability density on G such that $F_\alpha \simeq \rho^{-2\alpha-D}$. Then, the right hand side of the above inequality is comparable to

$$(1/2) \int |f(xh) - f(x)|^2 F_\alpha(h)\, dh\, dx,$$

which is nothing but the Dirichlet form associated with the convolution by the function F_α. Hence, together with Theorem VII.2.5, Proposition VII.3.4 allows us to deduce the decay of the convolution powers of $F_0 = |\Omega^3|^{-1} 1_{\Omega^3}$ from the decay of $F_\alpha^{(k)}$. The point is that F_α is spread out, whereas F_0 is compactly supported, and we are going to be able to study directly the decay of $F_\alpha^{(k)}$.

VII.4 Volume growth and polynomial decay of convolution powers

Let G be as above. The first part of this section shows that the decay of convolution powers of some nicely spread-out functions can be estimated by elementary means.

Let $(B_j)_{j \in \mathbf{N}^*}$ be a sequence of measurable subsets of G such that $\gamma_j = \xi(B_j) < +\infty$. Given a sequence of positive real numbers $\lambda = (\lambda_j)_{j \in \mathbf{N}^*}$ such that $\sum_{i=1}^{+\infty} \lambda_j = 1$, we will denote by F_λ the function defined by $F_\lambda = \sum_{j=1}^{\infty} \lambda_j \gamma_j^{-1} 1_{B_j}$. We will also set $s_j = \sum_{i=j}^{+\infty} \lambda_i$ and remark that $\int_G F_\lambda = s_1 = 1$.

VII.4.1 Lemma *There exists C such that, for all $k = 1, 2, \ldots,$*

$$\|F_\lambda^{(k)}\|_\infty \leq Ck \sum_{j=1}^{+\infty} \gamma_j^{-1}(s_j - s_{j+1}) e^{-ks_{j+1}}.$$

Proof Let us denote by ϕ_i the function $\lambda_i \gamma_i^{-1} 1_{B_i}$. Developing $F_\lambda^{(k)}$, we obtain

$$\|F_\lambda^{(k)}\|_\infty \leq \sum_{i_1, \ldots, i_k} \|\phi_{i_1} * \cdots * \phi_{i_k}\|_\infty.$$

We can estimate $\|\phi_{i_1} * \cdots * \phi_{i_k}\|_\infty$ by the product of the L^1 norms of $(k-1)$ of the ϕ_i and of the L^∞ norm of the remaining one. If we choose to take the

L^∞ norm of the ϕ_i with $i = \max\{i_1, ..., i_k\}$, we get

$$\|F_\lambda^{(k)}\|_\infty \leq \sum_{i_1,...,i_k} \gamma_{\sup\{i_1,...,i_k\}}^{-1} \prod_{\nu=1}^k \lambda_{i_\nu}$$

$$= \sum_{j=1}^\infty \gamma_j^{-1} \left((\lambda_1 + ... + \lambda_j)^k - (\lambda_1 + ... + \lambda_{j-1})^k \right)$$

$$\leq k \sum_{j=1}^\infty \gamma_j^{-1} \lambda_j (\lambda_1 + ... + \lambda_j)^{k-1} \leq ck \sum_{j=1}^\infty \gamma_j(s_j - s_{j+1})e^{-ks_{j+1}}.$$

Here, we used the fact that $(\lambda_1 + ... + \lambda_j)^k - (\lambda_1 + ... + \lambda_{j-1})^k$ is exactly the sum of all products $\prod_{\nu=1}^k \lambda_{i_\nu}$ with $i_\nu \leq j$ and at least one factor such that $i_\nu = j$.

VII.4.2 Lemma *Suppose that $(B_j)_{j\in\mathbb{N}^*}$ and $\lambda = (\lambda_j)_{j\in\mathbb{N}^*}$ are such that $\gamma_j \geq cs_j^{-\beta}$, $s_{j+1} \geq cs_j$, $\lambda_{j+1} \geq c\lambda_j$, for some $\beta > 0$ and $c > 0$. Then we have*

$$\|F_\lambda^{(k)}\|_\infty = O(k^{-\beta}).$$

Proof We deduce from VII.4.3 and the hypothesis that

$$\|F_\lambda^{(k)}\|_\infty \leq ck \sum_{j=1}^\infty s_{j+1}(s_j - s_{j+1})e^{-ks_{j+1}}.$$

The function $t \mapsto t^\beta e^{-kt}$ has exactly one maximum on $[0,1]$, obtained for $t_k = \beta/k$ and at which its value is $(\beta/ek)^\beta$. Let j_k be such that

$$s_{j_k}^\beta e^{-ks_{j_k}} = \max\{s_j^\beta e^{-ks_j}\};$$

it follows that

$$\sum_{j \geq j_k} s_j^\beta (s_{j-1} - s_j)e^{-ks_j} \leq (\beta/ek)^\beta s_{j_k-1}.$$

However, we certainly have either $s_{j_k} \leq t_k$ or $c^2 s_{j_k-1} < s_{j_k+1} < t_k < s_{j_k}$, and thus

$$\sum_{j \geq j_k} s_j^\beta (s_{j-1} - s_j)e^{-ks_j} \leq ck^{-\beta-1}.$$

We also know

$$\sum_{j < j_k} s_j^\beta (s_{j-1} - s_j)e^{-ks_j} \leq c \sum_{j < j_k} s_j^\beta (s_j - s_{j+1})e^{-ks_j}$$

$$\leq c \int_0^{+\infty} t^\beta e^{-kt}\, dt = C'k^{-\beta-1}.$$

We finally obtain

$$\|F_\lambda^{(k)}\|_\infty = O(k^{-\beta}).$$

Note that the hypothesis $\gamma_j \geq cs_j^{-\beta}$, $s_{j+1} \geq cs_j$, $\lambda_{j+1} \geq c\lambda_j$ implies somehow that F_λ is spread out with some regularity.

VII.4.3 Example Let $\lambda(\alpha)$ be the sequence $\lambda(\alpha) = (c(\alpha)j^{-1-\alpha})_{j\in\mathbf{N}^*}$, where $c(\alpha) = (\sum_{j=1}^{+\infty} j^{-1-\alpha})^{-1}$. With a slight change of notation, let us denote by F_α the function associated as above with the sequence $\lambda(\alpha)$. Notice first that $F_\alpha(h) \simeq \rho^{-2\alpha-D}(h)$. Then, if $\lambda_j \geq cj^n$, Lemma VII.4.2 gives $\|F_\alpha^{(k)}\|_\infty = O(k^{-n/\alpha})$.

Note that in some cases this result is sharp. To see this, let $G = \mathbf{R}^n$, $B_j = \{x \in \mathbf{R}^n \mid |x| \leq j\}$ and $\alpha = 1$. In that case, $\lambda_j = cj^n$. Moreover, we easily see that $F_1(x) \simeq (1+|x|)^{-n-1}$. Using VII.2.5, we see that $F_1^{(k)}$ should behave like the Poisson kernel $ck/(k^2+|x|^2)^{(n+1)/2}$, that is, should be $O(k^{-n})$. Indeed, Lemma VII.4.2 gives that $\|F_1^{(k)}\|_\infty = O(k^{-n})$. Till now, we have not used the fact that G is compactly generated.

We can now derive the behaviour of the convolution powers of $F_0 = |\Omega^3|^{-1} 1_{\Omega^3}$.

VII.4.4 Proposition *If* $V(n) \simeq n^D$, *then* $\|F_0^{(k)}\|_\infty = O(k^{-D/2})$.

Proof Let us apply VII.3.4 with $\alpha = \frac{1}{2}$. We obtain

$$\int (f(xh) - f(x))^2 \rho(h)^{-1-D} \, dh \, dx \leq C\|(I - T_0)^{1/4} f\|_2^2.$$

Let

$$F_1 = c(1) \sum_{j=1}^{+\infty} j^{-2} \gamma_j^{-1} 1_{B_j}$$

be the function introduced in Example VII.4.3, with $\gamma_j = V(j)$ and $B_j = \Omega^j$. Our hypothesis (i.e. $V(n) \simeq n^D$) implies by VII.4.3 that $\|F_1^{(k)}\|_\infty = O(k^{-D})$. But we also have $F_1(h) \simeq \rho(h)^{-1-D}$ and thus

$$\int |f(xh) - f(x)|^2 F_1(h) \, dh \, dx \leq C\|(I - T_0)^{1/4} f\|_2^2.$$

Applying VII.2.5, we obtain that the function F_0 satisfies

$$\|F_0^{(k)}\|_\infty = O(k^{-D/2}),$$

or equivalently that T_0 satisfies $(R_{D/2})$.

Consider now a probability density F on G, and define $\check{F}$ by $\check{F}(x) = F(x^{-1})$. Let T be the operator of right convolution by F; its adjoint is given by the operator of right convolution by $\check{F}$. It follows that

$$\|f\|_2^2 - \|Tf\|_2^2 = (f * (\delta - F * \check{F}), f).$$

Now, since $F * \check{F}$ is a symmetric function,

$$(f * (\delta - F * \check{F}), f) = \frac{1}{2} \int \int (f(xh) - f(x))^2 (F * \check{F})(h) \, dh \, dx.$$

To compare such a Dirichlet form with the one associated with F_0, we will need the

VII.4.5 Lemma *Let U be an open and generating neighbourhood of e in G, and K a compact subset of G. Then there exists $\nu_0 \in \mathbf{N}^*$ and $c > 0$ such that $1_U^{(\nu_0)} \geq c 1_K$.*

Proof One checks easily that the set

$$\{x \in G \mid \exists \nu_x \in \mathbf{N}^* \text{ such that } 1_U^{(\nu_x)}(x) > 0\}$$

is a subgroup of G that contains U; hence it is all of G. Since K is compact, and since $1_U^{(\mu)}(x) > 0$ as soon as $1_U^{(\nu)}(x) > 0$ and $\mu \geq \nu$, the conclusion follows from the continuity of $1_U^{(\nu)}$ for $\nu \geq 2$.

We can now give a proof of Theorems VII.1.1 and VII.2.2 under the additional assumption that $V(n) \simeq n^D$ for some $D > 0$. Let $F_1, ..., F_k, ...$ be as in VII.1.2, and $G_k = F_{\nu_0 k + 1} * \cdots * F_{\nu_0(k+1)}$, for some fixed ν_0 and $k \geq 1$. Using Lemma VII.4.5, choose ν_0 so that $1_U^{(\nu_0)} \geq c 1_{\Omega^3}$; since $F_k \geq \varepsilon 1_U$, it follows that $G_k \geq c' F_0$, hence that $G_k * \check{G}_k \geq c'' F_0^{(2)}$. Now

$$\|f\|_2^2 - \|T_0 f\|_2^2 = \left(f, (I - T_0^2) f \right)$$

$$= \frac{1}{2} \int \int (f(xh) - f(x))^2 F_0^{(2)}(h) \, dh \, dx$$

$$\leq \frac{1}{2c''} \int \int (f(xh) - f(x))^2 G_k * \check{G}_k(h) \, dh \, dx$$

$$= \frac{1}{c''} \left(\|f\|_2^2 - \|T_k f\|_2^2 \right),$$

where T_k is the operator of right convolution by G_k.

From Proposition VII.4.4, under the hypothesis $V(n) \simeq n^D$, we know that $\|F_0^{(k)}\|_\infty = O(k^{-D/2})$, hence that $\|T_0^k\|_{1 \to 2} = O(k^{-D/4})$. We can therefore write

$$\|f\|_2^2 = \|T_0^k f\|_2^2 + \sum_{j=0}^{k-1} \left(\|T_0^j f\|_2^2 - \|T_0^{j+1} f\|_2^2 \right)$$

$$\leq C_0 k^{-D/2} \|f\|_1^2 + k \left(\|f\|_2^2 - \|T_0 f\|_2^2 \right).$$

The inequality follows from

$$\|T_0^j f\|_2^2 - \|T_0^{j+1} f\|_2^2 = \| (I - T_0)^{1/2} T_0^j f\|_2^2$$

$$\leq \| (I - T_0)^{1/2} f\|_2^2 = \|f\|_2^2 - \|T_0 f\|_2^2.$$

Changing f to $T_j f$, $j \geq 1$, yields

$$\|T_j f\|_2^2 \leq C_0 k^{-D/2} \|f\|_1^2 + k \left(\|T_j f\|_2^2 - \|T_0 T_j f\|_2^2 \right).$$

Note that

$$\|T_j f\|_2^2 - \|T_0 T_j f\|_2^2 \leq C \|f\|_1^2,$$

since $\|G_j\|_\infty \leq C$. We can thus choose $k \in \mathbf{N}^*$ such that

$$k \leq \left\{ 2 C_0 C \|f\|_1^2 / \left(\|T_j f\|_2^2 - \|T_0 T_j f\|_2^2 \right) \right\}^{2D/(D+2)} < k+1,$$

and get

$$\|T_j f\|_2^{2+4/D} \leq C' \left(\|T_j f\|_2^2 - \|T_0 T_j f\|_2^2 \right) \|f\|_1^{4/D}.$$

Now remember that our starting point was

$$\|f\|_2^2 - \|T_0 f\|_2^2 \leq C \left(\|f\|_2^2 - \|T_k f\|_2^2 \right).$$

This yields

$$\|T_j f\|_2^{2+4/D} \leq C'' \left(\|T_j f\|_2^2 - \|T_k T_j f\|_2^2 \right) \|f\|_1^{4/D}.$$

Applying Lemma VII.2.6 to the sequence T_k of operators of right convolution by G_k, we get

$$\|G_1 * \dots * G_k\|_\infty = O(k^{-D/2}),$$

and thus, in addition,

$$\|F_1 * \dots * F_k\|_\infty = O(k^{-D/2}).$$

VII.5 The case of superpolynomial growth

In this section, we offer an alternative proof of the above results. This proof also yields optimal results in the case of superpolynomial growth. Moreover, this approach is technically simpler than the one we have just presented. However, we believe that the techniques of Section 4 are of intrinsic interest.

Let G be a unimodular compactly generated group, and Ω a compact generating neighbourhood of the origin in G. Let $V(n)$ be the Haar measure of Ω^n. For f a measurable function on G, define f_n by

$$f_n(x) = \frac{1}{V(n)} \int_{x \Omega^n} f(y)\, dy.$$

Write

$$\|f\|_2 \leq \|f - f_n\|_2 + \|f_n\|_2.$$

Thanks to Proposition VII.3.2, the first term can be estimated by

$$n \|(I - T_0)^{1/2} f\|_2,$$

where T_0 is the left convolution by $F_0 = |\Omega^3|^{-1}1_{\Omega^3}$. As for the second term, it is straightforward to check that $||f_n||_2 \leq V(n)^{-1/2}||f||_1$. Therefore, if we assume that $V(n) \geq C_V^{-1}n^D$, we obtain

$$||f||_2 \leq Cn||(I - T_0)^{1/2}f||_2 + \sqrt{C_V}n^{-D/2}||f||_1.$$

Now let S be a regularizing operator. Changing f to Sf in the above inequality yields

$$||Sf||_2 \leq Cn||(I - T_0)^{1/2}Sf||_2 + \sqrt{C_V}n^{-D/2}||f||_1.$$

Now $||(I - T_0)^{1/2}Sf||_2 \leq C_1||f||_1$, since S is regularizing.

Thus, since we can always assume $C_V \geq 1$, we can choose $n \in \mathbf{N}*$ to be the integral part of

$$\left(2C_1\sqrt{C_V}||f||_1/||(I - T_0)^{1/2}Sf||_2\right)^{2/(D+2)}.$$

This gives

$$||Sf||_2 \leq C''C_V^{1/(D+2)}||(I - T_0)^{1/2}Sf||_2^{D/(D+2)}||f||_1^{2/(D+2)},$$

hence

$$||Sf||_2^{2+4/D} \leq C''C_V^{2/D}||(I - T_0)^{1/2}Sf||_2^2||f||_1^{4/D}. \tag{1}$$

Note that C'' only depends on $||S||_{1\rightarrow1}, ||S||_{\infty\rightarrow\infty}$, and $||S||_{1\rightarrow\infty}$.

We are now ready to prove Theorem VII.1.2. Let $F_1, ..., F_k, ...$ be as in VII.1.2. Applying Lemma VII.4.5, we can obtain an integer ν_0 such that $1_U^{(\nu_0)} \geq c1_{\Omega^4}$. Hence, setting $G_j = F_{\nu_0 j+1} * ... * F_{\nu_0(j+1)}$, we have $G_j * \check{G}_j \geq c^2 1_{\Omega^4} * 1_{\Omega^4} \geq c'F_0$. Thus, denoting again by T_j the operator of left convolution by G_j, we have

$$||(I - T_0)^{\frac{1}{2}}f||_2^2 = \frac{1}{2}\int\int (f(xh) - f(x))^2 F_0(h)\,dh\,dx$$
$$\leq \frac{1}{2c'}\int\int (f(xh) - f(x))^2 G_j * \check{G}_j(h)\,dh\,dx$$
$$= \frac{1}{c'}\left(||f||_2^2 - ||T_jf||_2^2\right).$$

From this and inequality (1), we deduce that

$$||T_jf||_2^{2+4/D} \leq CC_V^{2/D}\left(||T_jf||_2^2 - ||T_kT_jf||_2^2\right)||f||_1^{4/D},$$

for all integers j, k. By Lemma VII.2.6, this yields $||G_1 * ... * G_k||_\infty \leq (C'D)^{D/2}C_V Dk^{-D/2}$. In particular,

$$||F_1 * ... * F_k|| = O(k^{-D/2}).$$

This ends the proof of the polynomial growth case in Theorems VII.1.1 and VII.1.2.

The superpolynomial case is obtained from the polynomial one by approximation. Assume indeed that $V(n) \geq ce^{cn^{\alpha}}$, for some $\alpha \in]0,1[$. By elementary estimates, this implies $V(n) \geq c_1(c_2\alpha/D)^{D/\alpha}n^D$, $\forall D > 0$, where c_1, c_2 depend only on c. Now, the above argument shows that

$$\|F_1 * \ldots * F_k\|_\infty \leq (C'D)^{D/2}c_1^{-1}(D/c_2\alpha)^{D/\alpha}k^{-D/2}, \quad \forall D > 0 \text{ and } k \geq 2.$$

Choosing $D = \varepsilon k^{\alpha/(\alpha+2)}$ for small enough ε, we obtain

$$\|F_1 * \ldots * F_k\|_\infty = O(e^{-c'k^{\alpha/(\alpha+2)}}).$$

This ends the proof of Theorems VII.1.1 and VII.1.2.

References and comments

The main results of this chapter and the overall strategy are due to Varopoulos [145], [155]. The presentation has however undergone several changes and substantial improvements since the original proofs. The abstract theory of Section VII.2, which is a discrete analogue of Chapter II, is taken from [35], [36], where some more results may be found. The paper [20] contains related results.

The comparison of Dirichlet forms in Section VII.3 is adapted from [145] and [109], [110]. The treatment of the case of polynomial volume growth given in Section VII.4 is taken from [109], [110], [35], [36]. In [154], [155] the approach of Section VII.4 is extended to the case of superpolynomial growth, with the help of the Log Sobolev type inequalities studied by E.B. Davies in [42].

The proof of Theorem VII.1.2 presented in Section VII.5 is adapted from an idea of D. Robinson as in [103]. Yet another proof of Theorem VII.1.1 (for symmetric F) is given in [63]. This paper contains further results concerning the case where $V(n) \simeq n^D$.

The bounds given by Theorem VII.1.1 are optimal (see [63] for the polynomial growth case, [2] and [8] for polycyclic groups having exponential growth).

Two-sided gaussian estimates for convolution powers when $V(n) \simeq n^D$ are derived in [63]. Previous results in the direction of gaussian upper bounds had been obtained in [134], [21], [20].

CHAPTER VIII

THE HEAT KERNEL ON UNIMODULAR LIE GROUPS

In this chapter, we shall link the local theory of Chapter V and the analysis of Chapter VII to estimate the heat kernel on a unimodular Lie group with respect to the volume growth. We treat Section 2 the case of the polynomial volume growth, and then derive parabolic Harnack inequalities. In Section 4, we give an alternative approach, which is also relevant in the case of exponential growth groups.

VIII.1 Preliminaries

In this chapter, G will be a connected unimodular Lie group, endowed with its Haar measure, and $\mathbf{X} = \{X_1, ..., X_k\}$ a Hörmander system of left invariant vector fields. As in Section III.4, we associate with $\mathbf{X}$ the Carnot-Carathéodory distance $(x, y) \in G \times G \to \rho(x, y)$ and denote by $\rho(x)$ the distance from the origin to $x \in G$; $V(t)$ will be the volume of the balls $B(x, t)$ centred at $x \in G$ and of radius $t > 0$ for that distance. By Proposition III.4.2, the behaviour of the distance ρ for sufficiently distant points essentially depends on G and not on $\mathbf{X}$. The same holds for the behaviour of the function $t \to V(t)$ for $t \geq 1$. By results recalled from VI.3.5, this behaviour is, in the Lie group setting, either of exponential type, or of polynomial type.

The sublaplacian associated with $\mathbf{X}$ is $\Delta = -\sum_{i=1}^{k} X_i^2$ and H_t, $t \geq 0$, is the symmetric Markov semigroup considered in II.5.1. Thanks to the left invariance of Δ, H_t admits a right convolution kernel h_t such that

$$H_t f(x) = \int_G h_t(y^{-1}x) f(y) \, dy = f * h_t(x).$$

The hypoellipticity of $\frac{\partial}{\partial t} + \Delta$, which follows from the Hörmander condition, implies that $h_t(x)$ is a C^∞ function of $(t, x) \in \mathbb{R}^{*+} \times G$. The gradient ∇ associated with $\mathbf{X}$ is given by $\nabla f = (X_1 f, ..., X_k f)$, and we put

$$\|\nabla f\|_p = \left(\int_G |\nabla f|^p \right)^{1/p}, \quad |\nabla f| = \left(\sum_{i=1}^{k} |X_i f|^2 \right)^{1/2}.$$

The analysis of the kernel h_t for small time was worked out in Chapter V and we shall use it further. We are now going to apply the ideas of Chapter VII to obtain the behaviour of h_t for large time. The following mean value inequality will be of fundamental importance in what follows.

VIII.1.1 Lemma *For every p, $1 \leq p \leq +\infty$, we have*

$$\|f_h - f\|_p \leq \rho(h) \|\nabla f\|_p, \quad \forall f \in C_0^\infty(G), \forall h \in G.$$

In particular (Δ, ρ) fulfills $(\mathcal{H}_2)$.

Proof Let $h \in G$ and $\gamma : [0, a] \to G$ be a path of $\mathcal{C}_X$ connecting e to h. In other words $\gamma(0) = e$, $\gamma(a) = h$, $\gamma'(t) = \sum_{i=1}^{k} a_i(t) X_i(\gamma(t))$ a.e., and $\sum_{i=1}^{k} a_i^2(t) \leq 1$. Then

$$|f(xh) - f(x)| \leq \int_0^a |\nabla f(x\gamma(t))| \, dt.$$

Hence, by integrating over G and using the translation invariance of the Haar measure,

$$\|f_h - f\|_p \leq a\|\nabla f\|_p.$$

It suffices then to minimize on a to obtain the desired conclusion. The case $p = 2$ gives $(\mathcal{H}_2)$.

VIII.2 Polynomial growth Lie groups

Throughout this section, we shall suppose that the volume growth of G is polynomial, and more precisely that there exists $D \in \mathbf{N}^*$ such that $V(t) \simeq t^D$, $t \geq 1$; we shall say that G has polynomial growth of order D. Recall that, for $0 \leq t \leq 1$, $V(t) \simeq t^d$, where $d \in \mathbf{N}$ is the local dimension of $(G, \mathbf{X})$, which was computed in Section V.1. We put $\Phi(h) = [V(\rho(h))]^{-1}$.

VIII.2.1 Proposition *For every $\alpha \in]0, 1[$, we have*

$$\int \left(\|f_h - f\|_2 / \rho^\alpha(h) \right)^2 \Phi(h) \, dh \leq C_\alpha \|\Delta^{\alpha/2} f\|_2^2, \quad \forall f \in C_0^\infty(G).$$

Proof The hypotheses of Proposition VII.3.2 are fulfilled: (Δ, ρ) satisfies $(\mathcal{H}_2)$ by Lemma VIII.1.1 and the fact that (ρ, Φ) satisfies $(\mathcal{H}_1)$ is an exercise left to the reader.

VIII.2.2 Proposition *If G has polynomial growth , then*

$$h_t(e) \leq C V(\sqrt{t})^{-1}, \quad \forall t > 0.$$

Proof The estimate for small time follows from Section V.4. We now have to show that

$$h_t(e) \leq C t^{-D/2}, \quad \forall t \geq 1.$$

Let us use the functions F_α introduced in VII.4.3, choosing

$$B_j = \{ x \in G \mid \rho(x) \leq j \},$$

hence $\gamma_j = V(j)$, $j \in \mathbf{N}^*$, so that

$$F_\alpha(h) = C_\alpha \sum_{j=1}^{+\infty} j^{-1-\alpha} V(j)^{-1} 1_{B_j}.$$

Since by hypothesis $\gamma_j = V(j) \geq Cj^D$, we have, via VII.4.2,

$$||F_{2\alpha}^{(k)}||_\infty < Ck^{-D/2\alpha}, \quad k \in \mathbf{N}^*.$$

Let T be the operator of convolution by ϕ and P_α the operator of convolution by $F_{2\alpha}$. Proposition VII.2.1 yields

$$||T(I - P_\alpha)^{-\frac{1}{2}} f||_{2D/(D-2\alpha)} \leq C||f||_2, \quad \forall f \in C_0^\infty(G),$$

i.e.

$$||Tf||_{2D/(D-2\alpha)} \leq C||(I - P_\alpha)^{1/2} f||_2.$$

Now

$$||(I - P_\alpha)^{1/2} f||_2 = \int ||f_h - f||_2^2 F_{2\alpha}(h) \, dh.$$

On the other hand, $\Phi(h) \simeq \rho(h)^{-D}$ if $\rho(h) \geq 1$, and therefore

$$F_{2\alpha}(h) \simeq \rho^{-2\alpha}(h)\Phi(h) \text{ if } \rho(h) \geq 1.$$

Finally

$$||Tf||_{2D/(D-2\alpha)}^2 \leq C \int (||f_h - f||_2/\rho^\alpha(h))^2 \, \Phi(h) \, dh.$$

Proposition VIII.2.1 then gives, for $0 < \alpha < 1$,

$$||Tf||_{2D/(D-2\alpha)} \leq C||\Delta^{\alpha/2} f||_2, \quad \forall f \in C_0^\infty(G).$$

By choosing $\phi = h_1$, and using II.4.3, the conclusion follows.

From the above theorem and Chapter II we deduce, as in IV.6.1 (see also Remark IV.6.3),

VIII.2.3 Theorem   *If G has polynomial growth of order D and if the local dimension of $(G, \mathbf{X})$ is $d \leq D$, then for every $1 < p < +\infty$, $\alpha > 0$, $n \in [d, D]$ such that $\alpha p < n$, $\Delta^{-\alpha/2}$ is bounded from L^p to $L^{p/(n-\alpha p)}$. Moreover, $\Delta^{-\alpha/2}$ is bounded from L^1 to $L^{n/(n-\alpha),\infty}$ and from L^1 to $L^{n/(n-\alpha)}$ if $n \in]d, D[$.*

Notice the important particular case of the above theorem:

$$\forall n \in [d, D] \cap]2, +\infty[, \ ||f||_{2n/(n-2)} \leq C||\nabla f||_2, \forall f \in C_0^\infty(G).$$

We are now going to derive the upper gaussian estimates of the kernel h_t.

VIII.2.4 Theorem   *If G has polynomial growth, for every $m \in \mathbf{N}$ and $\varepsilon > 0$, there exists $C_{m,\varepsilon} > 0$ such that*

$$\left|\left(\frac{\partial}{\partial t}\right)^m h_t(x)\right| \leq C_{m,\varepsilon} t^{-m} V(\sqrt{t})^{-1} \exp(-\rho^2(x)/(4+\varepsilon)t), \ \forall x \in G, \forall t > 0.$$

Proof Let us first consider the case $2 < d \leq D$. We shall use the perturbation technique explained in II.5.7 and II.5.8. Let $\lambda \in \mathbb{R}$ and $\phi \in C_0^\infty(G)$ be real and such that $|\nabla \phi|^2 = \sum_{i=1}^k |X_i \phi|^2 \leq 1$. Set

$$\psi = e^{\lambda \phi}, \quad B = \psi^{-1} \Delta \psi, \quad \text{and} \quad S_t = \psi^{-1} H_t \psi,$$

ψ denoting at the same time the function and the multiplication operator. It is clear that S_t is a semigroup generated by $-B$ which acts in the sense of Chapter II on the L^p spaces, $1 \leq p \leq +\infty$. Moreover, for $f \in C_0^\infty(G)$, we have

$$(Bf, f) = \int \nabla(\psi f) \cdot \nabla(\psi^{-1} \bar{f})$$
$$= \|\nabla f\|_2^2 - \lambda^2 \int |\nabla \phi|^2 |f|^2 - \lambda \int \nabla \phi \cdot (f \nabla \bar{f} - \bar{f} \nabla f).$$

The last term being purely imaginary, we get, since $|\nabla \phi| \leq 1$,

$$\text{Re}\,(Bf, f) \geq \|\nabla f\|_2^2 - \lambda^2 \|f\|_2^2 \geq -\lambda^2 \|f\|_2^2. \tag{1}$$

Moreover, for f positive, a simple computation shows that

$$(Bf, f^{p-1}) = (\Delta f, f^{p-1}) - \lambda^2 \int |\nabla \phi|^2 |f|^p - \lambda(p-2) \int \nabla \phi \cdot f^{p-1} \nabla f.$$

Notice that

$$\int f^{p-1} |\nabla f| \leq \|f\|_p^{p/2} \left(\int f^{p-2} |\nabla f|^2 \right)^{\frac{1}{2}}$$
$$\leq \frac{1}{2} \left(\varepsilon \|f\|_p^p + \varepsilon(p-1)^{-1} (\Delta f, f^{p-1}) \right), \quad \forall \varepsilon > 0.$$

This gives

$$(Bf, f^{p-1}) \geq [1 - (p-2)(\lambda)/2\varepsilon(p-1)](\Delta f, f^{p-1}) - (\lambda^2 + \varepsilon|\lambda|(p-2)/2)\|f\|_p^p,$$

and, for $\varepsilon = |\lambda|(p-2)/(p-1)$, we get

$$(Bf, f^{p-1}) \geq \frac{1}{2}(\Delta f, f^{p-1}) - p\lambda^2 \|f\|_p^p. \tag{2}$$

Note that (1) and (2) are also fulfilled by $B^* = \psi \Delta \psi^{-1}$, the dual operator of B. Thanks to Theorem VIII.2.3, we can apply Corollary II.5.8 with $A = \Delta$, $\alpha = \lambda^2$, and $n \in [d, D]$. We therefore obtain

$$\|S_t\|_{1 \to \infty} \leq C t^{-n/2} (1 + \lambda^2 t)^{n/2} e^{\lambda^2 t}, \quad \forall t > 0.$$

The semigroup S_t has a kernel given by

$$p_t(x, y) = e^{-\lambda \phi(x)} h_t(y^{-1} x) e^{\lambda \phi(y)}.$$

Thus
$$p_t(x,e) \leq \|S_t\|_{1\to\infty} \leq Ct^{-n/2}(1+\lambda^2 t)^{n/2}e^{\lambda^2 t},$$
which finally gives
$$h_t(x) \leq Ct^{-n/2}(1+\lambda^2 t)^{n/2}\exp[\lambda^2 t + \lambda(\phi(x)-\phi(e))].$$
Having fixed x and t, we may choose for ϕ a compactly supported approximation of $\zeta \mapsto \rho(\zeta)$ such that $\phi(e) \simeq 0$, $\phi(x) \simeq \rho(x)$ and $|\nabla\phi| \leq 1$, since $|\rho(z) - \rho(y)| \leq \rho(y^{-1}z)$. Passing to the limit, we get
$$h_t(x) \leq Ct^{-n/2}(1+\lambda^2 t)^{n/2}\exp(\lambda^2 t + \lambda\rho(x)),$$
and choosing $\lambda = -\rho(x)/2t$,
$$h_t(x) \leq Ct^{-n/2}(1+\rho^2(x)/t)^{n/2}\exp(-\rho^2(x)/4t), \quad \forall t>0, \forall x \in G.$$
Theorem VIII.2.4 follows for $m=0$. To obtain the estimates of the derivatives $(\frac{\partial}{\partial t})^m h_t$, it suffices to compute $\operatorname{Re}(e^{i\varepsilon}Bf, f)$ for $\varepsilon < \frac{\pi}{2}$, and to apply Remark II.5.9.

Now, if $D < d$ or $d < 2$, consider the group $G_0 = H^N$, where H is the Heisenberg group of Section IV.2. G_0 is endowed with the Hörmander system obtained by taking on each factor H the fields $X, Y, Z = [X, Y]$ considered in Section IV.2. We have then, in the notation of Chapter IV, $d(G_0) = 3N$ and $D(G_0) = 4N$. Let us take $G' = G \times G_0$, endowed with the Hörmander system obtained by putting together those of G and of G_0. It follows, if $x \in G$, $x_0 \in G_0$ and in the obvious notation, that
$$h_t^{G'}((x,x_0)) = h_t^G(x)h_t^{G_0}(x_0).$$
On the other hand, for N big enough, we have $V_{G'}(t) \simeq t^{d'}$, $0 \leq t \leq 1$, $V_{G'}(t) \simeq t^{D'}$, $t \geq 1$ where $2 < d' = d + 3N \leq D' = D + 4N$.

Finally the estimates of $h_t^{G'}$ will give those of $h_t = h_t^G$, since the upper and lower estimates of $h_t^{G_0}$ are at our disposal, and since $h_t^G(x) = h_t^{G'}((x,e))/h_t^{G_0}(e)$. Similarly, the upper estimates of the derivatives of h_t follow from the upper estimates of $h_t^{G'}$, $h_t^{G_0}$ and their derivatives and the lower estimate of $h_t^{G_0}$.

We are now going to derive the gaussian estimates of first order space derivatives of the kernel h_t. We first need an intermediate result.

VIII.2.5 Lemma   *For every $m \in \mathbf{N}$ and $p \geq 1$, there exists C_m such that*
$$\left\|\left(\frac{\partial}{\partial t}\right)^m h_t(.)\exp(\alpha\rho(.))\right\|_p \leq C_m t^{-m}V(\sqrt{t})^{-1/p'}e^{C_m\alpha^2 t}, \quad \alpha > 0, t > 0.$$

Proof   We give it for $m=0$. It suffices to prove the inequality for $p = +\infty$, which is an elementary exercise, and for $p = 1$. To this end, we first check that there exists a constant $C > 0$ such that
$$h_t(x)\exp(\alpha\rho(x)) \leq Ce^{C\alpha^2 t}V(\sqrt{t})^{-1}\exp(-\rho^2(x)/Ct).$$

We compute then

$$\int \exp(-\rho^2(x)/Ct)\, dx = \int_{\rho^2(x)\leq t} + \sum_{i=1}^{+\infty} \int_{2^{i-1}t\leq \rho^2(x)\leq 2^i t}$$
$$\leq C[V(\sqrt{t}) + \sum_{i=1}^{+\infty} V(\sqrt{2^i t})\exp(-2^i/C)] \leq CV(\sqrt{t}),$$

which gives the result.

VIII.2.6 Corollary *The semigroup $(H_t)_{t\geq 0}$ is bounded analytic on L^1 when G has polynomial growth.*

This result is immediate: $||\frac{\partial h_t}{\partial t}||_1 \leq C/t$ follows from Lemma VIII.2.5. We shall not use it.

VIII.2.7 Theorem *If G has polynomial growth, then for $i \in \{1,...,k\}$ and $m \in \mathbf{N}$ we have*

$$\left|\left(\frac{\partial}{\partial t}\right)^m X_i h_t(x)\right| \leq C_m t^{-m-1/2} V(\sqrt{t})^{-1} \exp(-\rho^2(x)/C_m t),$$

for all $t > 0$ and $x \in G$.

Proof We give it again for $m = 0$. Write

$$X_i h_t(x) = \int h_{t/2}(y) X_i h_{t/2}(y^{-1}x)\, dy.$$

Then, for $\alpha > 0$,

$$|e^{\alpha\rho(x)} X_i h_t(x)| \leq \int h_{t/2}(h) e^{\alpha\rho(y)} |X_i h_{t/2}(y^{-1}x)| e^{\alpha\rho(y^{-1}x)}\, dy,$$

using the triangle inequality. It follows that

$$|e^{\alpha\rho} X_i h_t| \leq ||h_{t/2} e^{\alpha\rho}||_2 ||X_i h_{t/2} e^{\alpha\rho}||_2.$$

The first factor of the right hand side is controlled, thanks to VIII.2.5, by $V(\sqrt{t})^{-\frac{1}{2}} e^{C\alpha t}$. To estimate the second factor, consider the function $\phi \in C_0^\infty$ such that $0 \leq \phi$ and $|\nabla\phi| \leq 1$. It is possible to approximate ρ by such functions, since $|\rho(x) - \rho(y)| \leq \rho(y^{-1}x)$. Then write

$$||X_i h_{t/2} e^{\alpha\rho}||_2 = \lim_n ||X_i h_{t/2} e^{\alpha\phi_n}||_2.$$

Now, if ϕ is as above,

$$||X_i h_{t/2} e^{\alpha\phi}||_2^2 = -\int h_{t/2}(x) X_i^2 h_{t/2}(x) e^{2\alpha\phi(x)}\, dx$$
$$- 2\alpha \int h_{t/2}(x) X_i h_{t/2}(x) X_i\phi(x) e^{2\alpha\phi(x)}\, dx,$$

hence

$$\sum_{i=1}^{k}||X_i h_{t/2}e^{\alpha\phi}||_2^2 = \int h_{t/2}(x)\Delta h_{t/2}(x)e^{2\alpha\phi(x)}\,dx$$

$$- 2\alpha\sum_{i=1}^{k}\int h_{t/2}(x)X_i h_{t/2}(x)X_i\phi(x)e^{2\alpha\phi(x)}\,dx.$$

In particular

$$||X_i h_{t/2}e^{\alpha\phi}||_2^2 \le ||h_{t/2}e^{2\alpha\phi}||_2 \left(||\Delta^{\frac{1}{2}}h_{t/2}||_2 + 2\alpha C||\nabla h_{t/2}||_2\right).$$

But since $-\Delta h_t = \frac{\partial h_t}{\partial t}$ and $||\nabla h_t||_2^2 = ||\Delta^{\frac{1}{2}}h_t||_2^2$, it is possible to control all the terms of the right hand side thanks to Lemma VIII.2.5.

Putting together all these estimates, we find

$$|e^{\alpha\rho(x)}X_i h_t(x)| \le Ct^{-\frac{1}{2}}V(\sqrt{t})^{-1}e^{C\alpha^2 t}, \quad \alpha > 0, t > 0.$$

In other words

$$|X_i h_t(x)| \le Ct^{-1/2}V(\sqrt{t})^{-1}\exp(C\alpha^2 t - \alpha\rho(x)).$$

If we optimize, with x and t fixed, by choosing $\alpha = \rho(x)/2Ct$, we get

$$|X_i h_t(x)| \le Ct^{-1/2}V(\sqrt{t})^{-1}\exp(-\rho^2(x)/4Ct).$$

This ends the proof.

VIII.2.8 Lemma   *If G has polynomial volume growth,*

$$\int_{\rho(x)\ge r} h_t(x)\,dx \le Ce^{-r^2/Ct}, \quad r > 0, t > 0.$$

Proof   Write

$$\int_{\rho(x)\ge r} h_t(x)\,dx = \sum_{i=0}^{+\infty}\int_{2^i r\le\rho(r)\le 2^{i+1}r} h_t(x)$$

$$\le CV(\sqrt{t})^{-1}\sum_{i=0}^{+\infty}V(2^i r)e^{-4^i r^2/Ct}.$$

If $t \le 1$ and $r \le 1$, denote by i_0 the smallest integer such that $2^{i_0}r \ge 1$, then $V(2^i r) \simeq (2^i r)^d$ if $i \le i_0$ and $V(2^i r) \simeq (2^i r)^D$ if $i \ge i_0$. If $d \le D$, then $(2^i r)^D \le 2^{iD}r^d$, whereas, if $d \ge D$, $(2^i r)^D \le (2^i r)^d$. In both cases we get

$$\int_{\rho(x)\ge r} h_t(x)\,dx \le C\left(\frac{r}{\sqrt{t}}\right)^d e^{-r^2/Ct}\sum_{i=0}^{+\infty}2^{i(d+1)}e^{-2^i/C} \le C'e^{-r^2/Ct}.$$

The cases $t \leq 1$ and $r \geq 1$, $t \geq 1$ and $r \leq 1$, $t \geq 1$ and $r \geq 1$ can be treated in the same way.

From Theorems VIII.2.4, VIII.2.7 and the above lemma, we are now going to deduce

VIII.2.9 Theorem   *If G has polynomial volume growth, there exist $C, c > 0$ such that*

$$cV(\sqrt{t})^{-1}\exp(-C\rho^2(x)/t) \leq h_t(x) \leq CV(\sqrt{t})^{-1}\exp(-c\rho^2(x)/t),$$

for all $t > 0$ and $x \in G$.

Proof   The upper estimate comes from VIII.2.4. Concerning the lower estimate, first notice that

$$h_t(e) \geq CV(\sqrt{t})^{-1}, \quad t > 0. \tag{3}$$

Indeed, Lemma VIII.2.8 implies

$$\int_{\rho(x) \geq At} h_t(x)\, dx \leq Ce^{-A^2/C} \leq 1/2,$$

for large enough A. It follows that

$$1/2 \leq \int_{\rho(x) \leq At} h_t(x)\, dx \leq \sup_x h_t(x)V(\sqrt{t}) = h_t(e)V(\sqrt{t}),$$

hence (3).

Then, using Theorem VIII.2.7, we shall deduce that there exists $a > 0$ such that

$$h_t(x) \geq CV(\sqrt{t})^{-1}, \quad \text{for} \quad \rho^2(x) \leq at. \tag{4}$$

Indeed,

$$|h_t(x) - h_t(e)| \leq \rho(x)\sup\{|X_i h_t(y)| \mid i \in \{1, ..., k\}, \rho(y) \leq \rho(x)\}.$$

From VIII.2.7, if $\rho^2(y) \leq \rho^2(x) \leq at$,

$$|X_i h_t(y)| \leq Ct^{-1/2}V(\sqrt{t})^{-1} \leq Ca^{1/2}\rho^{-1}(x)V(\sqrt{t})^{-1}.$$

Therefore

$$h_t(x) \leq h_t(e) - a^{\frac{1}{2}}CV(\sqrt{t})^{-1}.$$

Together with (3), this gives (4), for a small enough.

We still have to see that (4) implies the lower estimate of Theorem VIII.2.9. We can now suppose $\rho^2(x) > at$. For every $N \in \mathbf{N}^*$ we then have

$$h_t(x) = [(H_{t/N})^{N-1}h_{t/N}](x)$$
$$\geq \int_{S(x,N)} h_{t/N}(x_1^{-1}x)h_{t/N}(x_2^{-1}x_1)\cdots h_{t/N}(x_{N-1})\, dx_1 \cdots dx_{N-1}, \tag{5}$$

where $S(x, N) = \{(x_1, ..., x_{N-1}) \in G^{N-1} \mid \rho(x_i^{-1} y_i) \leq \rho(x)/N\}$ with $y_0 = x, y_1, ..., y_N = e$ a fixed sequence of points such that $\rho(y_{i+1}^{-1} y_i) \leq 2\rho(x)/N$, $i = 0, ..., N - 1$. Then, for every $(x_1, ..., x_{N-1}) \in S(x, N)$, $\rho(x_{i+1}^{-1} x_i) \leq 4\rho(x)/N$, $i = 0, ..., N-1$; we take $x_0 = x$, $x_N = e$. Choose now $N \in \mathbf{N}^*$ such that $N - 1 \leq 16\rho^2(x)/at < N$, so that for every $(x_1, ..., x_{N-1}) \in S(x, N)$, $\rho^2(x_{i+1}^{-1} x_i) \leq at/N$. We then use (4) and (5) to get

$$h_t(x) \geq [CV(\sqrt{t/N})]^{-N} [V(\rho(x)/N)]^{N-1}.$$

Now since $N \approx \rho^2(x)/t$,

$$[CV(\sqrt{t/N})]^{-N} [V(\rho(x)/N)]^{N-1} \geq C^{-N} V(\sqrt{t})^{-1}$$
$$\geq CV(\sqrt{t})^{-1} \exp(-c\rho^2(x)/t).$$

This ends the proof of VIII.2.9.

The optimal results for the Sobolev inequalities are another application of VIII.2.7. Let us consider the Sobolev and Hölder seminorms $\|f\|_{p,\alpha}$ and Λ_α defined at the end of Section IV.7.

VIII.2.10 Theorem   *If G has polynomial growth of order D, then*

(i) $\|f\|_{n/(n-1)} \leq C\|\nabla f\|_1, \forall f \in C_0^\infty(G)$ *if and only if* $d \leq D$ *and* $n \in [d, D]$;

(ii)   *for every* $1 \leq p < +\infty$, $n \in [d, D]$, $\alpha \in \mathbf{N}^*$ *such that* $\alpha p < n$,

$$\|f\|_{pn/(n-\alpha p)} \leq C\|f\|_{p,\alpha}, \quad \forall f \in C_0^\infty(G);$$

(iii)   *for every* $1 \leq p < +\infty$, $n \in [d, D]$, $\alpha \in \mathbf{N}^*$ *such that* $\alpha p > n$ *and* $\alpha - n/p \notin \mathbf{N}$, *then*

$$\Lambda_{\alpha-n/p}(f) \leq C\|f\|_{p,\alpha}, \quad \forall f \in C_0^\infty(G).$$

Proof   (ii) follows easily from (i). To obtain (i), we apply IV.7.1. The estimate on $\|h_t\|_\infty$ is given by VIII.2.2, and the one on $\|X_i h_t\|_1$ follows from VIII.2.7. The necessity of $n \in [d, D]$ is obtained as in IV.7.2. The proof of (iii) follows the same lines as in IV.7.4. The restriction on α comes from the fact that we only have estimates on the first order space derivatives of the kernel h_t.

VIII.3 Harnack inequalities for polynomial growth groups

The upper and lower gaussian estimates of the kernel h_t open the way to a Harnack theorem in the setting of polynomial growth groups.

To this end, we are first going to study the kernel of the semigroup associated with the equation $(\frac{\partial}{\partial t} + \Delta)u = 0$ with Dirichlet boundary condition on the ball of centre x and radius R. By translation invariance it suffices of course to work with $x = e$, which we shall do from now on.

Let $B(R) = \{x \in G \mid \rho(x) < R\}$. Let H_t^R be the symmetric submarkovian semigroup associated with the closed Dirichlet form:

$$Q_R(f,f) = \int_{B(R)} |\nabla f|^2, \quad f \in \mathcal{D}_R,$$

where $\mathcal{D}_R \subset L^2(B(R))$ is the closure of $C_0^\infty(B(R))$ under the norm $\|f\|_2 + Q_R^{\frac{1}{2}}(f)$.

The semigroup H_t^R has a kernel

$$(t, x, y) \mapsto h_t^R(x, y) \in C^\infty(\mathbb{R}^{*+} \times B(R) \times B(R))$$

such that

$$H_t^R f(x) = \int_{B(R)} h_t^R(x, y) f(y)\, dy, \quad f \in C_0^\infty(B(R)).$$

Of course, this kernel is also the fundamental solution of the equation

$$(\frac{\partial}{\partial t} + \Delta)u = 0$$

with Dirichlet boundary condition in $\mathbb{R}^{+*} \times B(R)$. The analysis that we are going to carry on relies on these two classical facts:

VIII.3.1 Lemma   *For every positive solution u of $(\frac{\partial}{\partial t} + \Delta)u = 0$ in $]a, b[\times B(R)$, $a < s < t < b$, we have*

$$u(t, x) \geq \int_{B(R)} h_{t-s}^R(x, y) u(s, y)\, dy.$$

VIII.3.2 Lemma   *For all $y \in B(R)$ and $t > 0$, there exists a positive measure of total mass smaller than 1, $\mu_{y,t}$, supported on $[0, t] \times \partial B(R)$, such that*

$$h_t^R(x, y) = h_t(y^{-1}x) - \int_{[0,t] \times \partial B(R)} h_s(z^{-1}x) \mu_{y,t}(\, ds \times dz).$$

These two lemmas are classical applications of the maximum principle or, in probabilistic language, of the strong Markov principle for the associated diffusion.

VIII.3.3 Theorem   *Given a group G having polynomial growth and $0 < \delta < 1$, there exists a constant $C > 0$ such that for every $R > 0$:*
(i) $h_t^R(x, y) \leq CV(\sqrt{t})^{-1} \exp(-\rho^2(x, y)/Ct), \quad \forall t > 0, \forall (x, y) \in B(R)^2;$
(ii) $[CV(\sqrt{t})]^{-1} \exp(-C\rho^2(x, y)/t) \leq h_t^R(x, y),$
   for every $(t, x, y) \in]0, R^2[\times B(\delta R) \times B(\delta R).$

Proof   The first assertion is a simple consequence of Lemma VIII.3.2 and of the upper estimate of the kernel h_t given in VIII.2.4.

The proof of the second assertion is more difficult. First notice that the lower estimate of h_t given in VIII.2.10 and Lemma VIII.3.2 yield

$$h_t^R(x,y) \geq [CV(\sqrt{t})]^{-1} \exp(-C\rho^2(x,y)/t)$$
$$- C \sup_{0<s<t} V(\sqrt{s})^{-1} \exp[-(1-\delta^2)R^2/Cs]$$

for $x \in B(\delta R)$, $y \in B(R)$ and $t > 0$.

It follows that there exists $r > 0$ depending on C and δ but not on R such that

$$h_t^R(x,y) \geq [2CV(\sqrt{t})]^{-1} \exp(-C\rho^2(x,y)/t)$$
$$\text{for } x \in B(\delta R), y \in B(R), \rho(x,y) \leq rR \text{ and } t \in [0, (rR)^2]. \tag{1}$$

Consider $0 < \delta' < \delta < 1$, $x \in B(\delta'R)$, $y \in B(\delta'R)$, $\rho(x,y) > rR$, $t \in [0, (rR)^2]$. We can connect x to y while staying in $B(\delta'R)$ (for example by passing through the origin) by an admissible path of length smaller than $2R$.

Then let y_i, $i \in \{0, ..., n\}$ be some points chosen on this path so that $y_0 = x$, $y_n = y$, $\rho(y_i, y_{i+1}) \leq 2R/n$. Let B_i be the ball $B(y_i, rR/3)$ and let us choose $n \in [6/r, (6/r)+1]$ so that $2R/n < rR/3$. If $z_i \in B_i$, we then have $\rho(z_i, z_{i+1}) < rR$, which ensures by (1) that

$$h_{t/n}^R(z_i, z_{i+1}) \geq [2CV(\sqrt{t/n})]^{-1} \exp(-Cn(rR)^2/t). \tag{2}$$

Moreover, as soon as $(\delta-\delta') < r/3$, $B(\delta R)\cap B_i$ contains $B(y_i, (\delta-\delta')R)$ hence $|B(\delta R) \cap B_i| > V((\delta - \delta')R)$. This last inequality (2), and the semigroup property, used as in VIII.2.10, give

$$h_t^R(x,y) \geq [2CV(\sqrt{t/n})]^{-n} \exp(-Cn^2(rR)^2/t)V((\delta - \delta')R)^{n-1}$$
$$\geq [AV(\sqrt{t})]^{-1} \exp(-A\rho^2(x,y)/t),$$

which is the announced result. It remains then to study the case where $x \in B(\delta'R)$, $y \in B(\delta'R)$ and $t \in [(rR)^2, R^2]$. Let us then choose $n \in]r^{-2}, r^{-2}+1]$, so that $t/n < (rR)^2$; the semigroup property and the above then yield

$$h_t^R(x,y) \geq [V(\delta'R)]^{n-1}[AV(\sqrt{t/n})]^{-n} \exp[-An^2(\delta'R)^2/(rR)^2]$$
$$\geq [A'V(\sqrt{t})]^{-1}.$$

This suffices to conclude the proof of Theorem VIII.3.3.

From Lemma VIII.3.1 and Theorem VIII.3.3, we deduce

VIII.3.4 Lemma   *Given $0 < \alpha < 1$, $0 < \delta < 1$, there exists $0 < \varepsilon < 1$ such that for every $R > 0$, $s \in \mathbb{R}$ and every positive solution of $(\frac{\partial}{\partial t} + \Delta)u = 0$ in $[s - R^2, s] \times \overline{B(R)}$, we have*

$$u(t,x) \geq \varepsilon V(\delta R)^{-1} \int_{B(\delta R)} u(s - R^2, y) \, dy, \quad \forall(t,x) \in [s - \alpha R^2, s] \times \overline{B(\delta R)}.$$

Given $(s, x) \in R \times G$, $\rho > 0$ and $u \in C^\infty([s - \rho^2, s] \times \overline{B(x, \rho)})$, let us set

$$\text{Osc}(u, s, x, \rho) = \sup\{|u(t, y) - u(t', y')| \mid (t, y), (t, y') \in [s - \rho^2, s] \times \overline{B(x, \rho)}\}.$$

VIII.3.5 Lemma   *Given $0 < \delta < 1$, there exists $0 < a < 1$ such that*

$$\text{Osc}(u, s, x, \delta\rho) \leq a\text{Osc}(u, s, x, \rho)$$

as soon as $(s, x) \in \mathbb{R} \times G$, $u \in C^\infty([s - \rho^2, s] \times B(s, \rho))$ and $(\frac{\partial}{\partial t} + \Delta)u = 0$ in $]s - \rho^2, s[\times B(x, \rho)$, $\rho > 0$.

Proof   For $0 < r < \rho$, put

$$M(r) = \max\{u(t, y) \mid (t, y) \in [s - r^2, s] \times \overline{B(x, r)}\}$$
$$m(r) = \min\{u(t, y) \mid (t, y) \in [s - r^2, s] \times \overline{B(x, r)}\}.$$

Let $A = \{y \in B(x, \delta\rho) \mid u(s - \rho^2, y) \geq (M(r) + m(r))/2\}$. Suppose that $|A| \geq |B(x, \delta\rho)|/2$; then for $(t, y) \in [s - (\delta\rho)^2, s] \times B(x, \delta\rho)$, Lemma VI.3.4 gives

$$u(t, y) - m(\rho) \geq \varepsilon V(\delta R)^{-1} \int_{B(x, \delta R)} [u(s - \rho^2, z) - m(r)]\, dz$$
$$\geq \varepsilon(M(\rho) - m(\rho))/4.$$

It follows that $m(\delta\rho) - m(\rho) \geq \varepsilon(M(\rho) - m(\rho))/4$, hence

$$M(\delta\rho) - m(\delta\rho) \leq (1 - \varepsilon/4)(M(\rho) - m(\rho)).$$

In the case where $|A| < |B(x, \delta\rho)|/2$, it suffices to consider $B(x, \delta\rho)\backslash A$ and $M(r) - u$ to obtain the same result, which proves Lemma VIII.3.5.

VIII.3.6 Theorem   *Given a group G having polynomial growth and $0 < \alpha < \beta < 1$, $0 < \delta < 1$, there exists a constant C such that for all $(s, x) \in \mathbb{R} \times G$, all $R > 0$ and every positive solution $u \in C^\infty([s - R^2, s] \times \overline{B(x, R)})$ of $(\frac{\partial}{\partial t} + \Delta)u = 0$ in $]s - R^2, s[\times B(x, R)$, we have*

$$u(t, y) \leq Cu(s, x); \quad (t, y) \in [s - \beta R^2, s - \alpha R^2] \times \overline{B(x, \delta R)}.$$

VIII.3.7 Corollary   *Given a group G having polynomial growth, any solution u of $(\frac{\partial}{\partial t} + \Delta)u = 0$ in $\mathbb{R} \times G$ which is bounded below is constant. In particular, any solution u of $\Delta u = 0$ in G which is bounded below is constant.*

Proof of the Corollary   Assume, without loss of generality, that u is positive. Then Theorem VIII.3.6 shows that u is bounded, and VIII.3.5 implies that u is constant. Indeed, by iterating VIII.3.5, one gets

$$\text{Osc}(u, s, x, 1) \leq a^n\text{Osc}(u, s, x, \delta^{-n}) \leq a^n 2 \sup u.$$

It suffices then to let n go to infinity.

Proof of Theorem VIII.3.6 We shall suppose as above that $x = e$ but also that $s = 0$. Thus u is a positive solution of $(\frac{\partial}{\partial t} + \Delta)u = 0$ in $] - R^2, 0[\times B(R)$. We may suppose $u(0,0) = 1$, and Lemma VIII.3.4 gives

$$1 = u(0,0) \geq \varepsilon V(\delta' R)^{-1} \int_{B(\delta', R)} u(t, z)\, dz \geq \lambda |S(t, \lambda, R)|/AV(R),$$

for every $t \in] - R^2, -\alpha R^2[$, $0 < \delta' < 1$, $\lambda > 1$, where

$$S(t, \lambda, R) = \{x \in B(\delta' R) \mid u(t, x) > \lambda\}.$$

Therefore we have, for $t \in] - R^2, \alpha R^2[$ and every $\lambda > 0$,

$$|S(t, \lambda, R)| \leq AV(R)\lambda^{-1}.$$

Suppose that $(t, y) \in] - R^2, -\alpha R^2[\times S(t, \lambda, R)$ and that λ is big enough to ensure that, if $T = T(\lambda)$ is defined by $V(T) = 2AV(R)/\sigma\lambda$, where $\sigma = (1 - a)/2$, a being the constant which appears in Lemma VIII.3.5, we have $t - (2T)^2 \in] - R^2, -\alpha R^2[$ and $B(y, 2T) \subset B(\delta' R)$. Then

$$|B(y, T)| = V(T) = 2AV(R)/\sigma\lambda > |S(t, \sigma\lambda, R)|,$$

therefore there exists $z \in B(y, T)$ such that $u(t, z) < \sigma\lambda$. This implies, using Lemma VIII.3.5,

$$\mathrm{Osc}(u, t, y, 2T) \geq a^{-1}\mathrm{Osc}(u, t, y, T) \geq a^{-1}|u(t, y) - u(t, z)|$$
$$\geq a^{-1}(\lambda - \sigma\lambda) = \lambda(1 + a)/2a = b\lambda,$$

where $b > 1$. Thus there exists $(t', y') \in [t - (2T)^2, t] \times B(y, 2T)$ such that

$$u(t', y') \geq \mathrm{Osc}(u, t, y, 2T) \geq \lambda b.$$

Let us fix $\delta < \delta' < 1$ and let $\lambda > 0$ be such that there exists $(t_0, y_0) \in] - \beta R^2, -\alpha R^2[\times B(\delta R)$ with $u(t_0, y_0) \geq \lambda$. By the preceding argument we can construct a sequence of points (t_n, y_n) such that

$$\begin{cases} u(t_n, y_n) \geq b^n \lambda \\ t_n \in [t_0 - \sum_{k=1}^{n-1}(2T_k)^2, t_0], \quad y_n \in B(\delta + \sum_{k=1}^{n-1} T_k), \end{cases}$$

where T_k is defined by $V(T_k) = 2AV(R)/\sigma\lambda b^{k-1}$. To be able to pass from step n to step $n + 1$ in this construction we must only make sure that

$$t_0 - \sum_{k=1}^{n}(2T_k)^2 > -R^2 \quad \text{and} \quad \delta R + 2\sum_{k=1}^{n} T_k < \delta' R.$$

We shall have finally completed the proof of Theorem VIII.3.6 if we find a $\Lambda > 0$, independent from R such that, for every $\lambda > \Lambda$,

$$-\beta R^2 - \sum_{k=1}^{+\infty} (2T_k)^2 > -R^2 \text{ and } \delta R + 2\sum_{k=1}^{+\infty} T_k < \delta' R. \tag{3}$$

Indeed, in that case, the existence of (t_0, y_0) in $] - \beta R^2, -\alpha R^2[\times B(\delta R)$ such that $u(t_0, x_0) > \Lambda$, would imply that u is unbounded in $[-R^2, 0] \times \overline{B}(R)$, which is absurd. To prove the existence of Λ, notice that if $R \leq 1$, $T_k \leq cR(\lambda^{-1}b^{-k+1})^{1/d}$, hence (3) reduces to

$$-\beta - 4c^2 \lambda^{-2/d} b^{2/d} / (b^{2/d} - 1) > -1$$

and $\delta + 2c\lambda^{-1/d} b^{1/d} / (b^{1/d} - 1) < \delta'$, which proves the existence of Λ in that case. If $R \geq 1$, let N be the first integer such that

$$\frac{2AV(R)}{\sigma \lambda b^{N-1}} \leq 1;$$

then for $k \leq N$, $T_k \leq CR(\lambda^{-1}b^{-k+1})^{1/D}$ and $\sum_{k=1}^{N}(2T_k)^2 \leq c'\lambda^{-2/D}R^2$, $\sum_{k=1}^{N} T_k \leq C'\lambda^{-1/D}R$. On the other hand, if $k > N$, we have $T_k \leq CR^{D/d}(\lambda^{-1}b^{-k+1})^{-1/d} \leq 1$ and therefore:

$$\begin{cases} \text{if } d \leq D, \quad T_k \leq CR(\lambda^{-1}b^{-k+1})^{-1/d} \\ \text{if } d > D, \quad T_k \leq (CR^{D/d}(\lambda^{-1}b^{-k+1})^{-1/d})^{-d/D} = CR(\lambda^{-1}b^{-k+1})^{-1/D}. \end{cases}$$

In any case, we finally obtain the reduction of (3) to $-\beta - C\lambda^{-\alpha} > -1$ and $\delta + C'\lambda^{-\alpha'} < \delta'$ for some positive constants c, c', α, α' independent of R, which ends the proof of the Theorem.

VIII.4 Exponential growth Lie groups

At this point, we want to turn our attention to Lie groups having an exponential volume growth. Our first goal is to obtain a sharp estimate for the decay of the maximum of the heat kernel $h_t(e)$ for large t. The approach presented in Section VII.5 can easily be adapted to the continuous setting. However, we will take this opportunity to present yet another technique, quite different in spirit from the above developments. Indeed, the argument below is not based on an equivalence between the decay of the heat kernel and some functional inequality involving the Dirichlet form.

VIII.4.1 Theorem *Assume that G has exponential volume growth. Then there exist $c, C > 0$ such that*

$$h_t(e) \leq Ce^{-ct^{1/3}}, \forall t \geq 1.$$

Proof For $x \in G$, $s, t > 0$, write

$$h_{t+s}(x) - h_{t+s}(e) = \int_G h_{t/2}(y) \left[h_{t/2+s}(xy) - h_{t/2+s}(y) \right] dy,$$

hence, by Cauchy-Schwarz,

$$|h_{t+s}(x) - h_{t+s}(e)| \leq \|h_{t/2}\|_2 \|(h_{t/2+s}(x.) - h_{t/2+s}(.)\|_2.$$

Lemma VIII.1.1 gives

$$|h_{t+s}(x) - h_{t+s}(e)| \leq \rho(x)\|h_{t/2}\|_2 \|\nabla h_{t/2+s}\|_2.$$

Now $\|\nabla h_{t/2+s}\|_2 = \|\Delta^{1/2} h_{t/2+s}\|_2 \leq Cs^{-1/2}\|h_{t/2}\|_2$ (this estimate may be seen as a consequence of the analyticity of the heat semigroup), and $\|h_{t/2}\|_2^2 = h_t(e)$. Finally

$$|h_{t+s}(x) - h_{t+s}(e)| \leq Cs^{-1/2}\rho(x)h_{t/2}(e),$$

and in particular

$$h_{t+s}(x) \geq h_{t+s}(e) - Cs^{-1/2}\rho(x)h_{t/2}(e).$$

Set now $\gamma(t,s) = h_{t+s}(e)/2Cs^{-1/2}h_t(e)$. For all x such that $\rho(x) \leq \gamma(t,s)$ the above inequality yields

$$h_{t+s}(x) \geq \frac{1}{2}h_{t+s}(e).$$

By integrating over the ball $B(e, \gamma(t,s))$, we get

$$h_{t+s}(e) \leq 2V(\gamma(t,s)). \tag{1}$$

For $0 < s \leq t$, consider the $[t/s]$ points $t + is \in [t, 2t]$, $i = 0, ..., [t/s]$. Then, either there exists an $i_0 \in \{0, ..., [t/s] - 1\}$ such that

$$h_{t+(i_0+1)s}(e)/h_{t+i_0s}(e) \geq 1/2,$$

or, $\forall i \in \{0, ..., [t/s] - 1\}$,

$$h_{t+(i+1)s}(e)/h_{t+is}(e) \leq 1/2.$$

In the first case, $\gamma(t + i_0s, s) \geq s^{1/2}/4C$, and (1) yields

$$h_{2t}(e) \leq h_{t+(i_0+1)s}(e) \leq 2V(s^{1/2}/4C)^{-1}.$$

In the second case,

$$h_{2t}(e) \leq h_{t+[t/s]s}(e) \leq \left[\prod_{i=0}^{[t/s]-1} h_{t+(i+1)s}(e)h_{t+is}^{-1}(e)\right]h_t(e) \leq 2^{-[t/s]}h_t(e).$$

In any case, we have proved the following inequality:

$$h_{2t}(e) \leq \max\{2V(s^{1/2}/4C)^{-1}, 2^{-[t/s]}h_t(e)\}, \tag{2}$$

for all $0 < s \leq t$.

Note that we have not used any information on the volume growth yet. Now, if $V(t) \geq C_0 e^{c_1 t}$ for $t \geq 1$, we get, for $1 \leq s \leq t$,

$$h_{2t}(e) \leq \max\{2C_0 e^{-c_1 s^{1/2}/4C}, h_1(e)2^{-[t/s]}\},$$

since $h_t(e) \leq h_1(e)$, $t \geq 1$. Choosing $s = t^{2/3}$ in the above yields

$$h_{2t}(e) \leq C e^{-ct^{1/3}}, \ t \geq 1.$$

This ends the proof of Theorem VIII.4.1.

Note that inequality (1) can be used to obtain $h_t(e) = O(t^{-D/2})$ in the case where $V(t) \geq t^D$, $t \geq 1$. However, the iterative argument above has to be modified then.

From Theorem VII.4.3, one gets, by adapting the techniques of Section 2,

VIII.4.2 Theorem *If G has exponential growth and if the local dimension of $(G, \mathbf{X})$ is d, then for every $1 < p < +\infty$ and $\alpha > 0$ such that $\alpha p < d$, $\Delta^{-\alpha/2}$ is bounded from L^p to L^q, for any $q \in]p, \frac{dp}{d-\alpha p}]$.*

VIII.4.3 Theorem *If G has exponential growth and if the local dimension of $(G, \mathbf{X})$ is d, then for every $n \geq d$ and $\varepsilon > 0$, there exists $C_{n,\varepsilon} > 0$ such that*

$$|h_t(x)| \leq C_{n,\varepsilon} t^{-n/2} \exp(-\rho^2(x)/(4+\varepsilon)t), \quad \forall x \in G, \forall t > 0.$$

One can also derive Sobolev inequalities in this case, but this is slightly more involved. One first shows, thanks to Theorem V.4.2, that $\|X_i h_t\|_1 \leq C(t \wedge 1)^{-1/2}$; Theorems V.4.3 and VIII.4.1 give the behaviour of $\|h_t\|_\infty$. An adaptation of IV.7.1 then yields

VIII.4.4 Theorem *If G has exponential growth and if the local dimension of $(G, \mathbf{X})$ is d, then for all $n \geq d$,*

$$\|f\|_{n/(n-1)} \leq C_n \|\nabla f\|_1, \forall f \in C_0^\infty(G).$$

VIII.4.5 Remark The results presented here take their full significance in the setting of *amenable* groups. Indeed, if G is non-amenable, not only has G exponential growth, but there exists a spectral gap , i.e. $\lambda > 0$ such that

$$\|H_t\|_{2 \to 2} \leq e^{-\lambda t}, \forall t > 0.$$

Moreover, the strong Sobolev inequality

$$\|f\|_1 \leq C\|\nabla f\|_1, \quad \forall f \in C_0^\infty(G)$$

holds. In fact, it is well-known that each of these two properties is equivalent to the non-amenability of G.

References and comments

The Sobolev inequalities, the gaussian upper bounds and the diagonal two-sided bounds for heat kernels on unimodular Lie groups first appeared in [142], [143], [149]. To obtain the gaussian factor in the estimates, essential use is made of the ideas of E.B. Davies [40]. The proof of Proposition VIII.2.2 is taken from [111], [112]. The kernel estimates are then derived as in [149]. The estimates of first order space derivatives as well as the lower bound on the heat kernel were proved in [111], where other applications can be found. As in the case of nilpotent groups, the upper estimate of the heat kernel can be slightly improved; see References and Comments to Chapter IV. We do not know whether the corresponding lower bound holds or not for general groups of polynomial growth. Bounds on the kernels of the semigroups $e^{-t\Delta^k}$, $k \in \mathbb{N}^*$, are obtained in [113], [114].

The parabolic Harnack inequality VIII.3.6 is taken from [111], [112]. Previous to that, the corresponding global Harnack inequality for "harmonic " functions was done in [147], where also a proof of the second part of Corollary VIII.3.7 was given. The connection between parabolic Harnack inequality and two-sided gaussian bounds has a long history; see [95], [91], [92], [93], [10], [76], [47]. G. Alexopoulos ([5], [7]) has been able to complement Theorem VII.3.6 with a Harnack inequality for first order space derivatives. His approach is original and involves ideas from homogenization theory. It also leads (see [5], [7]) to a proof of the continuity of the Riesz transforms $X_i\Delta^{-1/2}$ on the spaces $L^p(G)$, $1 < p < +\infty$ ([5], [7]). This can be restated as $\|\nabla f\|_p \simeq \|\Delta^{1/2}f\|_p$, $1 < p < +\infty$. Partial results had been obtained in [111], [112], [81].

In a different direction, it is shown in [115] that Theorems VIII.2.9 and VIII.3.6 extend to operators of the form $\mathcal{L} = \sum_{i,j=1}^k X_i a_{i,j} X_j$, where the matrix function $(a_{i,j})$ satisfies a uniform ellipticity condition.

It turns out that Theorem VIII.4.1 was proved independently, around the same time, by W. Hebisch [62], D. Robinson [103] and N. Varopoulos [155]. Previous results in this direction can be found in [141], [142], [143]. The ingenious proof given here is due to W. Hebisch. D. Robinson's proof is presented in Chapter VII, Section 5, in the context of convolution powers. Varopoulos's approach extends the ideas of Section VIII.2 and makes use of the Log Sobolev inequalities of [42]. For further results on Sobolev imbedding theorems, see [32], [33], [34], [111], [112], [78], [79], [80]. Sobolev imbedding theorems for *non*-compactly supported functions are considered in [6].

CHAPTER IX

SOBOLEV INEQUALITIES
ON NON-UNIMODULAR LIE GROUPS

In this chapter, we shall consider non-unimodular groups. One of the main motivations for this is that the non-compact symmetric spaces admit realizations as such groups (we are referring of course to the well-known Iwasawa "KAN" decomposition). We shall however *not* develop this point of view here.

As far as Sobolev inequalities are concerned, the results of this chapter seem to be satisfactory, since they answer the natural questions arising from the presence of a non-trivial modular function.

As far as the heat kernel is concerned, we will not touch the main difficulty, i.e. its behaviour for large time. The estimates that we give in Section 1 are only aimed at obtaining local Sobolev inequalities.

The main information at infinity is given in Section 2 by a version of Hardy's inequality. This is enough to derive all the Sobolev inequalities, except the one that involves the L^1 norm of the gradient; as usual, it is the strongest one.

For that inequality, the main thrust of our methods seems to be necessary, together with a decomposition of G as a semi-direct product of $\mathbb{R}$ and a unimodular group, given in Section 3.

IX.1 Local theory

Let G be a locally compact group. Denote by dx a right invariant Haar measure on G. The modular function of G is the function $m: G \to \mathbb{R}_+^*$ defined by

$$\int f(gx)\, dx = m(g) \int f(x)\, dx.$$

As a matter of fact, $I_g(f) = \int f(gx)\, dx$ is a right invariant Haar measure on G, and thus differs from dx by a multiplicative constant we can denote by $m(g)$. An important consequence of the definition is that $m(xy) = m(x)m(y)$, for all $x, y \in G^2$.

We may also characterize m by

$$\int f(x^{-1})\, dx = \int f(x)m(x)\, dx.$$

To this end, it suffices to notice that $m(x)\, dx$ is a left invariant Haar measure on G. We shall indeed set $d^\ell x = m(x)\, dx$.

From now on we shall suppose that G is a real connected Lie group, and that it is not unimodular, i.e. $m \not\equiv 1$.

We put

$$||\,|f|\,||_p = \left(\int |f|^p \, d^\ell x \right)^{\frac{1}{p}} = \left(\int |f|^p m(x)\, dx \right)^{\frac{1}{p}}.$$

The scalar product $\int f(x)\overline{g(x)}\,dx$ on $L^2(G,\,dx)$ will be denoted by (f,g), and the scalar product $\int f(x)\overline{g(x)}\,d^\ell x$ on $L^2(G,\,d^\ell x)$ will be denoted by $(f,g)_\ell$. Let $\mathbf{X}$ be a Hörmander system of left invariant vector fields on G. The distance associated with $\mathbf{X}$ is still denoted by $\rho(x,y) = \rho(y^{-1}x)$; we also set, as usual, $\Delta = -\sum_{i=1}^k X_i^2$, $\nabla f = (X_1 f, ..., X_k f)$.

Note that a left invariant vector field $\mathbf{X}$ is formally skew-adjoint for the scalar product of $L^2(G,\,dx)$:

$$\int Xf(x)g(x)\,dx = -\int f(x)Xg(x)\,dx, \quad f,g \in C_0^\infty(G).$$

Because of its multiplicativity, the modular function is an eigenfunction for any left invariant vector field X:

$$Xm(x) = \lim_{t \to 0} \frac{m(xe^{tX}) - m(x)}{t} = \left(\lim_{t \to 0} \frac{m(e^{tX}) - 1}{t} \right) m(x).$$

Denote by $V(t)$ the Haar measure of $B(t)$, the ball of radius t for the above distance, centred at the origin. Since $B(t) = B^{-1}(t)$, $V(t)$ does not depend on the choice of the right or left invariant measure. According to Section V.1, there exists an integer d such that $V(t) \simeq t^d$ for $0 < t < 1$.

For large time the growth of V is exponential. Indeed, there must exist $g \in B(1)$ such that $m(g) > 1$, otherwise m would equal 1 on $B(1)$, hence by multiplicativity, on all of G. This implies that $V(t) \geq e^{ct}$, $t \geq 1$, for some $c > 0$, since

$$V(n+1) = \int 1_{B(n+1)}(x)\,dx \geq \int 1_{B(1)}(g^n x)\,dx = [m(g)]^n \int_{B(1)} dx.$$

In the above context, it is natural to consider the two following Dirichlet forms:

$$Q_r(f) = \int |\nabla f|^2\,dx = \|\nabla f\|_2^2,$$

and

$$Q_\ell(f) = \int |\nabla f|^2\,d^\ell x = \|\,|\nabla f|\,\|_2^2.$$

Let us first consider Q_r; we have

$$Q_r(f) = (\Delta f, f), \quad f \in C_0^\infty(G).$$

Denote by $R_t = e^{-t\Delta}$ the symmetric submarkovian semigroup associated with Q_r and by $r_t(x,y) = r_t(y,x)$ the kernel of R_t on $(G,\,dx)$:

$$R_t f(x) = \int_G r_t(x,y)f(y)\,dy, \quad f \in C_0^\infty(G).$$

Note that r_t is not a convolution kernel. However, it is easy to check that

$$r_t(x,y) = r_t(e, x^{-1}y)m(x).$$

Let us now consider Q_ℓ. We have

$$Q_\ell(f) = |||\nabla f|||_2^2 = \sum_{i=1}^{k} \int |X_i f|^2 m(x)\, dx$$

$$= -\sum_{i=1}^{k} \int X_i^2 f(x)\overline{f(x)}m(x)\, dx - \sum_{i=1}^{k} \int X_i f(x)\overline{f(x)}X_i m(x)\, dx$$

$$= \int [(\Delta - \sum_{i=1}^{k} \lambda_i X_i)f(x)]\overline{f(x)}m(x)\, dx,$$

where the λ_i's are such that $X_i m(x) = \lambda_i m(x)$. Putting

$$A = \Delta - \sum_{i=1}^{k} \lambda_i X_i,$$

we get

$$Q_\ell(f) = (Af, f)_\ell, \quad f \in C_0^\infty(G).$$

Denote by $H_t = e^{-tA}$ the symmetric submarkovian semigroup associated with Q_ℓ. Since A is left invariant, H_t admits a convolution kernel $h_t(x)$ with respect to $d^\ell x$, such that

$$H_t f(x) = \int h_t(y^{-1}x)f(y)\, d^\ell y.$$

An elementary calculation yields the identity

$$Af = \Delta f - \sum_{i=1}^{k} \lambda_i X_i f = m^{-1/2}[\Delta + \lambda/4](m^{1/2}f),$$

with $\lambda = \sum_{i=1}^{k} \lambda_i^2$, or $\Delta m = -\lambda m$. It follows that

$$H_t f = e^{-\lambda t/4} m^{-1/2} R_t(m^{1/2}f).$$

The kernels h_t and r_t are therefore related by the formulae

$$r_t(x, y) = e^{\lambda t/4} h_t(y^{-1}x)m^{1/2}(x)m^{1/2}(y)$$

and

$$h_t(x) = e^{-\lambda t/4} m^{-1/2}(x)r_t(x, e).$$

We are going to estimate r_t first.

By left invariance of Δ, the function $u(s, z) = r_{t+2s}(xz, yz)$ is a solution of $\left(\frac{\partial}{\partial s} + \Delta\right)u = 0$ in $]0, +\infty[\times G$. Applying Theorem V.3.1 to u at e with $t_1 = 1/2$ and $t_2 = 1/2(1 + \varepsilon)$, we obtain

IX.1.1 Proposition *Given $\varepsilon > 0$, $J \in \mathcal{I}(k)$ and $m \in \mathbf{N}$, there exists C such that:*

$$\sup_{z \in B_{\sqrt{s}}} \left|\left(\frac{\partial}{\partial t}\right)^m X^J r_{t+s}(xz, yz)\right| \le C s^{-m-|J|/2} \inf_{z \in B_{\sqrt{s}}} r_{t+(1+\varepsilon)s}(xz, yz),$$

for all $x, y \in G$, $t > 0$, $0 < s \leq 1$.

Consider now the perturbed semigroup

$$T_t f = e^{-\alpha\phi} R_t(e^{\alpha\phi} f)$$

where $\phi \in C_0^\infty(G)$ is such that $|\nabla\phi| \leq 1$. We then have, as in IV.4.2,

$$T_t = e^{-tB}, Bf = e^{-\alpha\phi}\Delta(e^{\alpha\phi} f)$$

and

$$(Bf, f) = \sum_{i=1}^{k} \int X_i(e^{-\alpha\phi}\overline{f}) X_i(e^{\alpha\phi} f)\, dx.$$

It follows that

$$\mathrm{Re}\,(Bf, f) \geq \|\nabla f\|_2^2 - \alpha^2 \|f\|_2^2.$$

Let

$$\lambda_0 = \inf\left\{\frac{(\Delta f, f)}{\|f\|_2^2}, f \in C_0^\infty(G)\backslash\{0\}\right\}$$

be the bottom of the spectrum of Δ. We then have

$$\mathrm{Re}\,(Bf, f) \geq (\lambda_0 - \alpha^2)\|f\|_2^2,$$

hence

$$\|T_t f\|_{2\to 2} \leq e^{(-\lambda_0 + \alpha^2)t}.$$

Following again the argument of IV.4.2, we deduce that

$$\int_{\xi\in xB(\sqrt{s})}\int_{\zeta\in yB(\sqrt{s})} r_{t+(1+\varepsilon)s}(\xi, \zeta)e^{\alpha(\phi(\zeta)-\phi(\xi))}\, d\xi\, d\zeta$$

$$= \left(T_{t+(1+\varepsilon)s}1_{yB(\sqrt{s})}, 1_{xB(\sqrt{s})}\right)$$

$$\leq e^{(-\lambda_0 + \alpha^2)(t+(1+\varepsilon)s)}V(x, \sqrt{s})^{1/2}V(y, \sqrt{s})^{1/2},$$

where $V(x, \sqrt{s})^{1/2}$ is the volume of $xB(\sqrt{s})$ with respect to the right invariant measure dx and thus equals $m^{-1}(x)V(\sqrt{s})$. Thanks to Proposition IX.1.1, we can estimate from below the left hand side of the above inequality by

$$C^{-1}V^{1/2}(x, \sqrt{s})V^{1/2}(y, \sqrt{s})e^{\alpha(\phi(y)-\phi(x))-2|\alpha|\sqrt{s}}r_{t+s}(x, y),$$

whenever $0 < s \leq 1$. We get in this way

$$r_{t+s}(x, y) \leq Cm^{1/2}(x)m^{1/2}(y)V^{-1}(\sqrt{s})e^E,$$

where $E = (-\lambda_0 + \alpha^2)(t + (1 + \varepsilon)s) + \alpha(\phi(x) - \phi(y)) + 2|\alpha|\sqrt{s}$. Choose now ϕ so that $\phi(x) - \phi(y) = \rho(x, y)$ and $\alpha = -\rho(x, y)/2(t + s)$; this yields

$$E \leq -\lambda_0(t + s) - \rho^2(x, y)/4(t + s) + \rho(x, y)/\sqrt{t + s} + \varepsilon\rho^2(x, y)/4(t + s).$$

Taking $s = t$ if $0 < t \leq 1$, $s = 1$ if $t > 1$, we obtain

IX.1.2 Theorem *For any $\varepsilon \in]0, 1[$, there exists $C_\varepsilon > 0$ such that*

$$r_t(x, y) \leq C_\varepsilon m^{1/2}(x) m^{1/2}(y)(t \wedge 1)^{-d/2} e^{-\lambda_0 t - \rho^2(x,y)/(4+\varepsilon)t},$$

for all $t > 0$ and all $x, y \in G$. Moreover, for $0 < t \leq 1$,

$$r_t(x, y) \geq C m^{1/2}(x) m^{1/2}(y) t^{-d/2} e^{-C\rho^2(x,y)/t}.$$

The lower estimate follows from the Harnack estimate IX.1.1 along the lines of IV.4.3.

We can now derive estimates on h_t.

IX.1.3 Theorem *The kernel h_t satisfies:*
(i) $h_t(x) \leq C(t \wedge 1)^{-d/2} e^{-(\lambda_0 + \lambda/4)t - \rho^2(x)/(4+\varepsilon)t}$, $\forall t > 0$, $x \in G$;
(ii) $\forall m \in \mathbb{N}$, $J \in \mathcal{I}(k)$, $\||\left(\frac{\partial}{\partial t}\right)^m X^J h_t(x)|\|_1 \leq C t^{-m-|J|/2}$, $\forall t \in]0, 1]$;
(iii) $h_t(x) \geq C t^{-d/2} e^{-C\rho^2(x)/t}$, $\forall t \in]0, 1]$, $x \in G$.

Proof Properties (i) and (iii) follow from Theorem IX.1.2 through the formula giving h_t in terms of r_t. Property (ii) follows through the same formula from Proposition IX.1.1 and the fact that $\||h_t|\|_1 \leq 1$.

Local Sobolev inequalities follow from Theorem IX.1.3.

IX.1.4 Theorem *Let $n \geq d$. For every $p \in [1, n[$, we have*

$$\||f|\|_{pn/(n-p)} \leq C \left(\||\nabla f|\|_p + \||f|\|_p \right), \quad \forall f \in L^p(G, d^\ell x).$$

Proof As usual, it suffices to prove the theorem for $p = 1$. Write $A = \sum_{i=1}^k X_i^* X_i$ where $X_i^* = X_i - \lambda_i$. Put $D_i = (I + A)^{-1} X_i^*$, so that

$$\sum_{i=1}^k D_i X_i = (I + A)^{-1} A = I - (I + A)^{-1}.$$

We have

$$D_i = \int_0^{+\infty} e^{-t} H_t X_i^* \, dt$$

and

$$\||H_t X_i^* f|\|_1 = \||\int h_t(x^{-1}y) X_i^* f(y) \, d^\ell y|\|_1$$
$$= \||\int X_i h_t(x^{-1}y) f(y) \, d^\ell y|\|_1 \leq \||X_i h_t|\|_1 \||f|\|_1.$$

Therefore, according to IX.1.3 (ii),

$$\||H_t X_i^*|\|_{1 \to 1} \leq C t^{-\frac{1}{2}}, \quad 0 \leq t \leq 1,$$

and

$$|||H_t X_i^*|||_{1 \to \infty} \leq C_n t^{-\frac{1}{2} - n/2}, \quad 0 < t \leq 1, \quad n \geq d.$$

We then first obtain, as in IV.7.1,

$$\int_{\{|f| > \lambda\}} d^\ell x \leq C[\lambda^{-1}(|||\nabla f|||_1 + |||f|||_1)]^{n/(n-1)}, \quad \forall f \in C_0^\infty(G),$$

then, using the same method as in IV.7.1, the strong inequality

$$|||f|||_{n/(n-1)} \leq C\,(|||\nabla f|||_1 + |||f|||_1), \quad \forall f \in C_0^\infty(G).$$

This implies the desired conclusion.

IX.2 An inequality of Hardy and some consequences

Let $\overline{G} = \mathrm{Ker}\,(m) = \{g \in G \mid m(g) = 1\}$. $\overline{G}$ is a closed normal subgroup of G and $\overline{G}$ is unimodular. Let us denote by $d\xi$ the Haar measure of $\overline{G}$ and identify $G/\overline{G}$ with $\mathbb{R}$. The functions

$$\tilde{x} \mapsto \int_{\overline{G}} f(\xi \tilde{x}) \, d\xi \quad \text{and} \quad \tilde{x} \mapsto \int_{\overline{G}} f(\tilde{x}\xi) \, d\xi$$

only depend on the class x in $G/\overline{G}$ of $\tilde{x}$ in G and we have:

$$\int_G f(g) \, dg = \int_{\mathbb{R}} dx \int_{\overline{G}} f(\xi \tilde{x}) \, d\xi$$
$$\int_G f(g) \, d^\ell g = \int_{\mathbb{R}} dx \int_{\overline{G}} f(\tilde{x}\xi) \, d\xi.$$

Let X be a left invariant vector field on G such that $Xm \not\equiv 0$, in other words such that X does not belong to the Lie algebra of $\overline{G}$ and let $R = \{\exp tX \mid t \in \mathbb{R}\}$ be the one-parameter group it generates. R is a closed group and $\overline{G} \cap R = \{e\}$. Indeed for every $x \in R\backslash\{e\}$, we have $m(x) \neq 1$, hence $\overline{G} \cap R = \{e\}$. To prove the first assertion, consider a sequence $x_n = \exp(t_n X)$ that converges to x in G, and put $a = m(\exp X)$; then $\log m(x_n x_{n-1}^{-1}) = a(t_n - t_{n-1}) \to 0$, and since $\log m(x_n) \to \log m(x)$, we see there exists $t = \lim t_n$ such that $x = \exp(tX) \in R$.

From the properties of R and $\overline{G}$ it follows that G may be written as the semidirect product of $\overline{G}$ with R: $G \simeq \overline{G} \rtimes R$. The disintegration formulae recalled above may then simply be interpreted as Fubini formulae stating that the mappings:

$$\overline{G} \times R \ni (\overline{g}, r) \to g = \overline{g}r \in G,$$
$$\overline{G} \times R \ni (\overline{g}, r) \to g = r\overline{g} \in G$$

respectively map the measures in the following way:

$$d\overline{g} \times dr \to dg$$
$$d\overline{g} \times dr \to d^\ell g.$$

IX.2.1 Proposition   *Let X be a left invariant vector field such that $Xm \not\equiv 0$, then for every $\alpha \neq 0$ and $1 \leq p < +\infty$ we have*

$$\int |f(g)|^p m^\alpha(g)\, dg \leq C \int |Xf|^p m^\alpha(g)\, dg, \quad \forall f \in C_0^\infty(G).$$

Consequently,

$$\|m^\alpha f\|_p \leq C \|m^\alpha \nabla f\|_p, \quad \forall f \in C_0^\infty(G).$$

Proof   Let us recall the following inequality of Hardy, where $\lambda \neq 0$:

$$\int_{-\infty}^{+\infty} |\phi(t)|^p e^{\lambda t}\, dt \leq C(\lambda, p) \int_{-\infty}^{+\infty} |\frac{\mathrm{d}}{\mathrm{d}t}\phi(t)|^p e^{\lambda t}\, dt, \quad \forall \phi \in C_0^\infty(\mathbb{R}).$$

Let β be such that $m(x) = e^{\beta x}$ for $x \in R = \{\exp tx \mid t \in \mathbb{R}\}$; we identify $x = \exp tX \in R$ with $t \in \mathbb{R}$. We have, for every $\xi \in \overline{G}$, $\frac{\mathrm{d}}{\mathrm{d}x}f(\xi x) = Xf(\xi x)$, hence, for $\alpha \neq 0$,

$$\int_{\mathbb{R}} |f(\xi x)|^p e^{\alpha \beta x}\, dx \leq C \int_{\mathbb{R}} |Xf(\xi x)|^p e^{\alpha \beta x}\, dx.$$

But $e^{\alpha \beta x} m^\alpha(\xi) = m^\alpha(\xi x)$, and by integrating with respect to $m^\alpha(\xi)\, d\xi$ on $\overline{G}$, we get the desired conclusion. To obtain the second assertion of Proposition IX.2.1, note that, among the fields $X_1, ..., X_k$ on which the gradient ∇ is built, there always exist a field X_i such that $X_i m \not\equiv 0$.

IX.2.2 Proposition   *Let $1 \leq p < +\infty$. For every $f \in C_0^\infty(G)$, we have:*
(i) $\| |f| \|_p \leq C \| |\nabla f| \|_p$;
(ii) $\forall \beta, \gamma \in \mathbb{R}$ *such that* $\beta + \gamma \neq 0, \| |m^\beta(\nabla m^\gamma f)| \|_p \leq C \| |m^{\beta+\gamma} \nabla f| \|_p$.

Proof   To obtain (i), it suffices to apply the previous proposition with $\alpha = \beta = 1$. To show (ii), observe that $X_i(m^\gamma f) = \lambda_i \gamma m^\gamma f + m^\gamma X_i f$. The result then follows from IX.2.1.

IX.2.3 Theorem   *Let $d \leq n \leq +\infty$, $\beta \in \mathbb{R}^*$, $1 \leq p < n$, $q = np/(n-p)$ and α such that $1/q - \alpha = 1/p - \beta$, then*

$$\|m^\alpha f\|_q \leq C \|m^\beta \nabla f\|_p, \quad \forall f \in C_0^\infty(G).$$

Proof   According to IX.1.4, we have, for every $d \leq n < +\infty$ and $1 \leq p < n$,

$$\| |\phi| \|_{np/(n-p)} \leq C \left(\| |\nabla \phi| \|_p + \| |\phi| \|_p \right), \quad \forall \phi \in C_0^\infty(G).$$

Proposition IX.2.2 (i) thus gives

$$\| |\phi| \|_{np/(n-p)} \leq C \left(\| |\nabla \phi| \|_p \right), \quad \forall \phi \in C_0^\infty(G).$$

Let us then take $\phi = m^{\beta-1/p}f$, $\beta \neq 0$; we get, with the notation of the theorem,

$$\|m^{\beta-1/p+1/q}f\|_q \leq C\|m^{1/p}\nabla(m^{\beta-1/p}f)\|_p$$

and since $\alpha = \beta - 1/p + 1/q$, assertion (ii) of IX.2.2 implies that

$$\|m^\alpha f\|_q \leq C\|m^\beta \nabla f\|_p, \quad \forall f \in C_0^\infty(G).$$

IX.2.4 Proposition *Suppose that* $-\infty < \alpha, \beta < +\infty$ *and* $1 \leq p, q < +\infty$ *are such that*

$$\|m^\alpha f\|_q \leq C\|m^\beta \nabla f\|_p, \quad \forall f \in C_0^\infty(G).$$

Then $1/q-\alpha = 1/p-\beta$, *and there exists* $n \in [d, +\infty]$ *such that* $q = np/(n-p)$.

Proof The hypothesis of Proposition IX.2.4, applied to the function f_g defined by $f_g(x) = f(gx)$, yields

$$m^{\alpha-1/q}(g)\|m^\alpha f\|_q \leq Cm^{\beta-1/p}(g)\|m^\beta \nabla f\|_p.$$

Obviously, this is possible for every $g \in G$ only if $\alpha - 1/q = \beta - 1/p$. Let us test our hypothesis on the functions

$$f_t(x) = (t - \rho(x))_+ = \sup\{t - \rho(x), 0\}.$$

For $0 < t \leq 1$, we obtain, as in IV.7.2,

$$c_\alpha V(t/2)^{1/q} \leq c_\beta V(t)^{1/p}.$$

Since $V(t) \simeq t^d$ for $0 < t \leq 1$, it follows that

$$\frac{1}{q} \geq \frac{1}{p} - \frac{1}{d},$$

and there exists $n \geq d$ such that

$$\frac{1}{q} = \frac{1}{p} - \frac{1}{n}.$$

The conclusion we draw from IX.2.4 is that Theorem IX.2.3 is the best possible, except for the restriction $\beta \neq 0$. In other words, it remains to decide the validity of the inequalities

$$\|m^{-1/n}f\|_{pn/(n-p)} \leq C\|\nabla f\|_p, \quad \forall f \in C_0^\infty(G) \tag{1}$$

for all $n \geq d$ and $1 \leq p < +\infty$. Note that the inequality

$$\|m^\alpha f\|_q \leq C\|\nabla f\|_p, \quad \forall f \in C_0^\infty(G),$$

automatically implies, thanks to IX.2.2 (i),

$$\|m^{\alpha+\gamma}f\|_q \leq C_\gamma \|m^\gamma \nabla f\|_p, \quad \forall f \in C_0^\infty(G), \forall \gamma \in \mathbb{R}.$$

IX.2.5 Proposition *For $d \leq n < +\infty$ and $1 < p < n$, we have*

$$\|m^{-1/n}f\|_{pn/(n-p)} \leq C\|\nabla f\|_p, \quad \forall f \in C_0^\infty(G).$$

Proof Let $\nabla_\alpha f = m^{-\alpha}\nabla(m^\alpha f)$, $\alpha \in \mathbb{R}$. The inequality (1) is then equivalent to

$$\| |f| \|_{pn/(n-p)} \leq C\| |\nabla_{1/p}f| \|_p, \quad \forall f \in C_0^\infty(G). \tag{2}$$

As a matter of fact, (1) for $m^{1/p}f$, yields

$$\|m^{1/p-1/n}f\|_{pn/(n-p)} \leq C\|\nabla(m^{1/p}f)\|_p,$$

i.e. (2). The advantage of this reformulation is that we have

$$\nabla_\gamma(f^s) = sf^{s-1}\nabla_{\gamma/s}f.$$

Suppose that, for some $\gamma \in \mathbb{R}$, we have

$$\| |f| \|_{n/(n-1)} \leq C\| |\nabla_\gamma f| \|_1, \quad \forall f \in C_0^\infty(G). \tag{3}$$

Applying this inequality to f^s, $s > 1$, we obtain

$$\| |f| \|_{n/(n-1)}^s \leq Cs\| |f^{s-1}\nabla_{\gamma/s}f| \|_1, \quad \forall f \in C_0^\infty(G).$$

Let us put $r = sn/(n-1)(s-1)$, $1/r + 1/p = 1$, $q = m/(n-1)$, and use Hölder's inequality. We get

$$\| |f| \|_q^s \leq Cs\| |f| \|_q^{s-1}\| |\nabla_{\gamma/s}f| \|_p$$

hence

$$\| |f| \|_q \leq Cs\| |\nabla_{\gamma/s}f| \|_p, \quad \forall f \in C_0^\infty(G).$$

Finally, by fixing $1 < p < n$ and choosing $s = p(n-1)/(n-p) > 1$, we see that the inequality (3) implies

$$\| |f| \|_{pn/(n-p)} \leq C\| |\nabla_{\gamma/s}f| \|_p, \quad \forall f \in C_0^\infty(G).$$

To conclude the proof of Proposition IX.2.5, it suffices to notice that Theorem IX.2.4, applied to the function $m^\gamma f$, $\gamma \neq 1$, for $p = 1$, $\alpha = 1 - 1/n$, $q = n/(n-1)$, $\beta = 1 - \gamma \neq 0$, shows that

$$\| |f| \|_{n/(n-1)} \leq C_\gamma \| |\nabla_\gamma f| \|_1, \quad \forall f \in C_0^\infty(G), \gamma \neq 1.$$

It follows that, for $1 < p < n$ and $s = p(n-1)/(n-p)$,

$$\| |f| \|_{pn/(n-p)} \leq C_\gamma \| |\nabla_{\gamma/s}f| \|_p, \quad \forall f \in C_0^\infty(G), \gamma \neq 1.$$

For $\gamma = (n-p)/(n-1)$ this gives

$$|||f|||_{pn/(n-p)} \leq C|||\nabla_{1/p}f|||_p, \quad \forall f \in C_0^\infty(G)$$

i.e. nothing but (2) or (1), which is the desired conclusion.

IX.3 A Sobolev inequality again

In view of Propositions IX.2.4 and IX.2.5, the only Sobolev inequalities which remain to study are

$$||m^{-1/n}f||_{n/(n-1)} \leq C||\nabla f||_1, \quad \forall f \in C_0^\infty(G),$$

for $d \leq n$. In this section, we are going to prove their validity, by using an analysis which deals with the group

$$\overline{G} = \mathrm{Ker}(m) = \{y \in G \mid m(g) = 1\}.$$

We again denote by ρ the distance induced on $\overline{G}$ by the distance ρ on G. The ball centred at the origin and of radius t for this distance will be denoted by $\overline{B}(t)$, and $\overline{V}(t) = \int_{\overline{G}} 1_{\overline{B}(t)}(x)\,dx$ will be its volume with respect to the Haar measure on $\overline{G}$.

IX.3.1 Proposition $\overline{V}(t) \geq e^{ct}$ *for* $t \geq 1$.

Proof According to the disintegration formulae for Haar measures on G recalled at the beginning of Section IX.2, we have

$$V(t) = \int_G 1_{B(t)}(g)\,d^\ell g = \int_{\mathbf{R}} \int_{\overline{G}} 1_{B(t)}(\tilde{x}\xi)\,d\xi\,dx.$$

Now

$$\int_{\overline{G}} 1_{B(t)}(\tilde{x}\xi)\,d\xi = |\overline{G} \cap \tilde{x}^{-1}B(t)|_{\overline{G}},$$

where $|A|_{\overline{G}}$ is the measure of A in $\overline{G}$.

Let us first remark that $\overline{G} \cap \tilde{x}^{-1}B(t)$ is a set of diameter smaller than or equal to $2t$ in $\overline{G}$, which implies

$$|\overline{G} \cap \tilde{x}^{-1}B(t)|_{\overline{G}} \leq \overline{V}(2t), \quad \forall \tilde{x} \in G, \forall t > 0.$$

On the other hand, the distance ρ induces by projection a distance on $G/\overline{G} \simeq \mathbf{R}$ which is equivalent to the Euclidean distance. It follows that if $\tilde{x} \in G$ belongs to the class $x \in G/\overline{G} \simeq \mathbf{R}$, we have

$$|x| \geq at \Rightarrow \rho(\tilde{x}) \geq t, t > 0.$$

In particular there exists $a' > 0$ such that if $|x| \geq a't$, $\overline{G} \cap \tilde{x}^{-1}B(t) = \emptyset$ for every $\tilde{x}$ belonging to the class of x.

From that we deduce

$$V(t) \leq \int_{-a't}^{a't} \overline{V}(2t)\, dx = 2a't\overline{V}(2t), \quad t > 0,$$

which leads to the announced conclusion, since we know that $V(t) \geq e^{ct}$.

From Proposition IX.3.1 and from Section VII.3, we deduce

IX.3.2 Proposition *The function*

$$F = C\sum_{j=1}^{+\infty} j^{-3}\overline{V}(j)^{-1}\chi_j,$$

where χ_j is the characteristic function of $\overline{B}(j)$ and $c^{-1} = \sum_{j=1}^{+\infty} j^{-3}$, satisfies

$$\|F^{(k)}\|_{L^\infty(G)} \leq C_n k^{-n}, \quad k \in \mathbf{N}^*,$$

for every $n > 0$, and

$$\int_G F(\xi)\rho(\xi)\, d\xi < +\infty.$$

This yields

IX.3.3 Lemma *For every $\phi \in C_0^\infty(\overline{G})$ and $n > 1$, there exists c such that*

$$\|f * \phi\|_{L^{n/(n-1)}(\overline{G})} \leq c\|f * (\delta - F)\|_{L^1(\overline{G})}.$$

Proof Let T be the right convolution by F operator and S the one by ϕ; from VII.2.1 and IX.3.2 it follows that

$$\|(I - T)^{-1}Sf\|_{L^{n/(n-1),\infty}(\overline{G})} \leq c\|f\|_{L^1(\overline{G})},$$

for every $n > 1$. To conclude, we invoke Marcinkiewicz's theorem.

IX.3.4 Lemma *Let $f_h(g) = f(gh)$. Then, for every $f \in C_0^\infty(G)$ and $h \in G$,*

$$\|f_h - f\|_1 \leq C\rho(h)\|\nabla f\|_1$$

and thus, if F is the function introduced in IX.3.3,

$$\int_{\overline{G}} \|f_h - f\|_1 F(h)\, dh \leq C\|\nabla f\|_1.$$

Proof To show that $\|f_h - f\|_1 \leq C\rho(h)\|\nabla f\|_1$ we proceed as in VIII.1.1. The second assertion follows from the first one and from the fact that

$$\int_{\overline{G}} \rho(h)F(h)\, dh < +\infty.$$

Let us now fix a decomposition of G as a semi-direct product $G \simeq \overline{G} \rtimes R$, where $R = \{x = \exp tX \mid t \in \mathbb{R}\}$, for some left invariant field X such that $Xm \not\equiv 0$, as in Section IX.2.

Given a function $f \in C_0^\infty(G)$ let

$$\|f\|_{L^p L^q} = \left(\int_R \left(\int_{\overline{G}} |f(x\xi)|^q \, d\xi \right)^{p/q} dx \right)^{1/p},$$

so that $\|f\|_{L^p L^p} = \|\,|f|\,\|_p$.

In the next section, we shall always identify a function $\phi \in C_0^\infty(\overline{G})$ with the measure $\phi(\xi)\, d\xi$, supported on $\overline{G}$, so that

$$f * \phi(g) = \int_{\overline{G}} f(g\xi^{-1})\phi(\xi)\, d\xi$$

is just a usual convolution product on G. In the same way, we shall identify a function $\chi \in C_0^\infty(\mathbb{R})$ with the measure $\chi(x)\, dx$ supported by R so that

$$f * \chi(g) = \int_R f(gx^{-1})\chi(x)\, dx$$

is again just a convolution product on G.

IX.3.5 Lemma *Let $\phi \in C_0^\infty(\overline{G})$ and $n > 1$. Then*

$$\|f * \phi\|_{L^1 L^{n/(n-1)}} \le C\|\nabla(mf)\|_1.$$

Proof Set $_x f(\xi) = f(x\xi)$, $x \in R, \xi \in \overline{G}$ and note that $(_x f) * \phi(\xi) = {}_x(f * \phi)(\xi)$ for $x \in R$ and $\xi \in \overline{G}$.

According to IX.3.3, we thus have

$$\|_x(f * \phi)\|_{L^{n/(n-1)}(\overline{G})} \le C\|_x f * (\delta - F)\|_{L^1(\overline{G})}$$
$$\le C \int_{\overline{G}} \int_{\overline{G}} |f(x\xi h) - f(x\xi)|F(h)\, dh\, d\xi$$

and, integrating with respect to x,

$$\|f * \phi\|_{L^1 L^{n/(n-1)}} \le C \int_R dx \left(\int_{\overline{G}} \int_{\overline{G}} |f(x\xi h) - f(x\xi)|F(h)\, dh\, d\xi \right)$$
$$= C \int_{\overline{G}} \|(mf)_h - mf\|_1 F(h)\, dh \le C'\|\nabla(mf)\|_1,$$

where IX.3.4 is used for the last inequality.

IX.3.6 Lemma *Let $\chi \in C_0^\infty(\mathbb{R})$ and $1 \le p \le r$. Then*

$$\|f * \chi\|_{L^r L^q} \le C\|f\|_{L^p L^q}, \quad \forall f \in C_0^\infty(G).$$

Proof Let us recall the vector space version of Young's theorem. If B is a Banach space and $0 < 1/q = 1/p + 1/s - 1$, $\theta \in C_0^\infty(\mathbb{R})$ and $\Phi \in C_0^\infty(\mathbb{R}, B)$, one has

$$||\Phi * \theta||_{L^r(\mathbb{R}, B)} \leq ||\theta||_{L^s(\mathbb{R})} ||\Phi||_{L^p(\mathbb{R}, B)}.$$

Now take $B = L^q(\overline{G})$, and for $f \in C_0^\infty(G)$,

$$\Phi(x, \xi) = f(\xi x) m^{1/q}(x) = f(\xi x) m^{1/q}(\xi x), \quad x \in R, \xi \in \overline{G}.$$

Finally if $\chi \in C_0^\infty(\mathbb{R})$ let

$$\chi_1(x) = \chi(x) m^{1/q}(x).$$

We then have

$$\begin{aligned}
||f * \chi||_{L^r L^q}^r &= \int_R \left(\int_{\overline{G}} |f * \chi(x\xi)|^q \, d\xi \right)^{r/q} dx \\
&= \int_R \left(\int_{\overline{G}} (|f * \chi(x\xi)| m^{1/q}(x))^q \, d\xi \right)^{r/q} dx \\
&= ||\Phi * \chi_1||_{L^r(\mathbb{R}, B)}^r \\
&\leq ||\chi_1||_{L^s(\mathbb{R})}^r ||\Phi||_{L^p(\mathbb{R}, B)}^r = C ||f||_{L^p L^q}^r.
\end{aligned}$$

This proves IX.3.6.

IX.3.7 Theorem *Let* $\phi \in C_0^\infty(\overline{G})$, $\chi \in C_0^\infty(\mathbb{R})$ *and* $\psi(g) = \psi(\xi x) = \phi(\xi) \chi(x)$ *for* $g = \xi x \in G = \overline{G} \rtimes R$. *Then*

$$|| \, |f * \psi| \, ||_{n/(n-1)} \leq C ||\nabla(mf)||_1, \quad \forall f \in C_0^\infty(G).$$

Proof According to IX.3.6 and IX.3.5 we have, since $f * \psi = f * \phi * \chi$, for $q = n/(n-1)$,

$$||f * \psi||_{L^q L^q} \leq C ||f * \phi||_{L^1 L^q} \leq C ||\nabla(mf)||_1,$$

but

$$||f * \psi||_{L^q L^q} = || \, |f * \psi| \, ||_q,$$

and this gives the result.

IX.3.8 Lemma *Let* $\psi \in C_0^\infty(G)$ *be such that* $||\psi||_1 = 1$ *and* $\psi \geq 0$. *Then*

$$||f * (\delta - \psi)||_1 \leq C_\psi ||\nabla f||_1, \quad \forall f \in C_0^\infty(G).$$

Moreover, if X *is a left invariant vector field,*

$$||f * X\psi||_1 \leq C_\psi ||\nabla f||_1, \quad \forall f \in C_0^\infty(G).$$

Proof We have

$$\|f * (\delta - \psi)\|_1 = \int_G \left| \int_G [f(x) - f(xh^{-1})]\psi(h)\,dh \right| dx$$

$$\leq \int_G \int_G |f(x) - f(xh^{-1})|\,dx\psi(h)\,dh$$

$$\leq C \int_G \rho(h)\psi(h)\,dh \|\nabla f\|_1,$$

using IX.3.4.

Now for the second assertion. Write $X\psi = \psi^+ - \psi^-$, where $\psi^+ = \sup\{X\psi, 0\}$. Since $\int X\psi = 0$, we have $\int \psi^+ = \int \psi^- = \gamma$. Set $\tilde{\psi}^+ = \psi^+/\gamma$ and $\tilde{\psi}^- = \psi^-/\gamma$, so that

$$f * X\psi = \gamma[(f * \tilde{\psi}^+ - f) - (f * \tilde{\psi}^- - f)].$$

By using IX.3.4, we get

$$\|f * \tilde{\psi}^+ - f\|_1 \leq \int \|f_h - f\|_1 \tilde{\psi}^+(h)\,dh \leq C \left(\int \tilde{\psi}^+(h)\rho(h)\,dh \right) \|\nabla f\|_1,$$

as well as the same sequence of inequalities for $\tilde{\psi}^-$. This easily ends the proof.

IX.3.9 Theorem   *Let $1 \leq p < n$, $d \leq n < +\infty$ and $q = pn/(n-p)$. Then we have*

$$\|m^{-1/n}f\|_q \leq C\|\nabla f\|_p, \quad \forall f \in C_0^\infty(G),$$

and more generally for $\alpha \in \mathbb{R}$, $\beta \in \mathbb{R}$ such that $1/p - \beta = 1/q - \alpha$,

$$\|m^\alpha f\|_q \leq C\|m^\beta \nabla f\|_p, \quad \forall f \in C_0^\infty(G).$$

Proof As we noticed at the beginning of this section, the only inequalities which remain to be proved are, for $n \geq d$,

$$\|m^{-1/n}f\|_{n/(n-1)} \leq C\|\nabla f\|_1, \quad \forall f \in C_0^\infty(G),$$

or equivalently,

$$\| \, |f| \, \|_{n/(n-1)} \leq C\|\nabla(mf)\|_1, \quad \forall f \in C_0^\infty(G).$$

Let ψ be as in Theorem IX.3.7. We have

$$\| \, |f * \psi| \, \|_{n/(n-1)} \leq C\|\nabla(mf)\|_1, \quad \forall f \in C_0^\infty(G),$$

and thus we are left with proving that

$$\| \, |f * (\delta - \psi)| \, \|_{n/(n-1)} \leq C\|\nabla(mf)\|_1.$$

From Theorem IX.1.4 and from the fact that

$$|| \, |\nabla \psi| \, ||_1 \leq C \, (|| \, |\nabla(m\psi)| \, ||_1 + || \, |\psi| \, ||_1)$$

we deduce

$$|| \, |f * (\delta - \psi)| \, ||_{n/(n-1)} \leq$$
$$C \, (|| \, |\nabla(m\psi)| \, ||_1 + || \, |\nabla(m(f * \psi))| \, ||_1 + || \, |f * (\delta - \psi)| \, ||_1)$$

or, putting $mf = \tilde{f}$, $m\psi = \tilde{\psi}$,

$$|| \, |f * (\delta - \psi)| \, ||_{n/(n-1)} \leq C \left(|| \, |\nabla \tilde{f}| \, ||_1 + || \, |\nabla(\tilde{f} * \tilde{\psi})| \, ||_1 + || \, |\tilde{f} * (\delta - \tilde{\psi})| \, ||_1 \right).$$

Suppose that ψ is normalized in such a way that $||\tilde{\psi}||_1 = 1$. Lemma IX.3.8 then gives

$$||\tilde{f} * (\delta - \tilde{\psi})||_1 + || \, |\nabla(\tilde{f} * \tilde{\psi})| \, ||_1 \leq C_\psi || \, |\nabla \tilde{f}| \, ||_1,$$

and thus

$$|| \, |f * (\delta - \psi)| \, ||_{n/(n-1)} \leq C || \, |\nabla \tilde{f}| \, ||_1 = C || \, |\nabla(mf)| \, ||_1,$$

which ends the proof.

References and comments

This chapter is taken from [148], [150], [151]. For another point of view and other results of the same type on symmetric spaces, see the paper of N. Lohoué [77]. Note that the behaviour of the heat kernel on semi-simple groups has been determined by Bougerol [16], see also [80].

CHAPTER X

GEOMETRIC APPLICATIONS

In this chapter we shall give some applications of the theory we have developed to geometry and related topics. More often than not we shall do this without entering into the details and by assuming that the reader is familiar with the necessary background of the various problems that will be examined. Indeed, to develop all the background material would take us out of the scope of the book. The reader could treat this final chapter either as an amusement park or alternatively as a source of inspiration for further study and research. Some open problems will be considered at the end of this chapter.

X.1 Geometry of Lie groups and quasiregular maps

Let G be a connected (unimodular or not) Lie group. We can define an essentially unique "left" or "right" Riemannian structure on G.

Indeed, let us fix some basis of $T_e(G)$ the tangent space at $e \in G$. A Riemannian structure on G is then uniquely determined if we require that it is invariant under the left action of G and that the above basis is orthonormal. Two different bases give rise to quasi-isometric Riemannian structures on G.

Quite generally, for two Riemannian manifolds (M_i, g_i), $i = 1, 2$ we say that they are quasi-isometric, and we say that $\phi \colon M_1 \to M_2$ is a quasi-isometry, if ϕ is a diffeomorphism and if

$$C^{-1} g_1(X, X) \leq g_2(d\phi_x(X), d\phi_x(X)) \leq C g_1(X, X), \forall x \in M_1, X \in T_x(M_1).$$

Similarly we can define right invariant Riemannian structures on G, and it is easy to verify that the mapping $G \to G$, $x \mapsto x^{-1}$ is a quasi-isometry from any of the above left invariant structures to any of the right invariant structures.

A natural generalization of quasi-isometries are the quasiconformal maps or, even more generally, the quasiregular maps. For the general theory of these maps we shall refer the reader to the literature. In particular, a non-constant C^1-mapping

$$\phi \colon (M_1, g_1) \to (M_2, g_2)$$

between two Riemannian manifolds is called quasiregular if there exists $C > 0$ such that

$$|d\phi(x)|^n \leq C J(\phi)(x); \quad x \in M_1$$

where $n \geq 1$ is the topological dimension of *both* M_1 and M_2, $|d\phi|$ indicates the operator norm of $d\phi \colon TM_1 \to TM_2$ and $J(\phi)$ is the Jacobian of ϕ. The reader will find in the literature a more general definition that applies to mappings which are not necessarily C^1. A quasiregular map that is one-to-one is called quasiconformal. When ϕ is a local homomorphism, the above definition simply says that ϕ takes small balls in M_1 to small ellipsoids of uniformly bounded eccentricities ("small" means of course that their diameters tend to zero). The main interest of the above maps is that they are natural generalizations of holomorphic functions $z \mapsto f(z)$ of *one* complex variable ($z \in \Omega$) i.e. of conformal mappings from one region of the complex plane to another. Seen like this, it is no surprise that one of the main issues here are the "Picard" Theorems, in other words theorems that say that under such and such a condition there *does not* exist a quasiregular map from one particular Riemannian manifold to another (constant maps are not allowed, remember).

It is precisely this type of Picard theorem for Lie groups that we shall examine in the next section but before this we shall need to recall some definitions and well-known facts about quasiregular maps.

Let (M, g) be a Riemannian manifold of dim $M = n \geq 1$ and let $K \subset\subset \Omega \subset M$ where Ω is open and K is compact; such a pair (K, Ω) is called a condensor and we shall define its (conformal) capacity by

$$C(K, \Omega) = \inf \int_M |\nabla u|^n \, d\mathrm{Vol}_M$$

the infimum being taken over all $u \in C_0^\infty(\Omega)$ such that $u(x) \geq 1$, for all $x \in K$.

It is an easy matter to verify that when $\phi \colon M_1 \to M_2$ is an open quasiconformal mapping between two Riemannian manifolds then $C(\phi(K), \phi(\Omega))$ is equal (up to a constant factor that depends only on ϕ) to $C(K, \Omega)$. This is the reason why the above capacity is called conformal. Other capacities with $|\nabla u|^n$ replaced by some other power $|\nabla u|^p$ can also be considered but they are *not* conformal invariants.

Now let $\phi \colon M_1 \to M_2$ be a continuous quasiregular map between two Riemannian manifolds. The first fact that we need is that $(\phi(K), \phi(\Omega))$ is a condensor in M_2 if (K, Ω) is a condensor in M_1. This is a consequence of the *non-trivial* fact that quasiregular maps are always open. The other fact we shall need, which is even less obvious, is that

$$C(\phi(K), \phi(\Omega)) \leq A \, C(K, \Omega),$$

where $A > 0$ depends only on ϕ and not on (K, Ω).

X.2 Picard Theorems on Lie groups

In this section we shall address ourselves to the following problem:

Let G_1, G_2 be two connected Lie groups of the same topological dimension endowed with, say, their left invariant Riemannian structures. Let us assume that $\phi \colon G_1 \to G_2$ is a quasi-isometry or more generally a quasiconformal or even a quasiregular map. What can then be said about the groups? How close are they to being (algebraically) isomorphic?

Not much is known about the above interesting questions. In this section we shall simply show how they can be handled with the methods of this book in the special case when $G_1 = \mathbb{R}^n$. The approach is based on the following easy

X.2.1 Proposition   *Let G be a Lie group of topological dimension $n \geq 1$ and let us assume that there exists some $D > n$ and $C > 0$ such that $V(t) \geq Ct^D$ for all $t \geq 1$. Then for every condensor (K, Ω) in G with K of positive measure, we have $C(K, \Omega) > 0$.*

Indeed, the Sobolev inequalities of Chapter IX imply that, for every $u \in C_0^\infty(\Omega)$ with $u \geq 1$ on K, we have

$$||\nabla u||_n \geq C|K|^{1/p}$$

for some $C > 0$ and $p \geq 1$ depending only on G. The proposition follows.

The other fact that we shall need is classical and easy to prove: it says that the condensor in $\mathbb{R}^n$ formed by $K = B$ the closed unit ball and $\Omega = \mathbb{R}^n$ has zero capacity.

An easy proof can be given by observing that we can make the infimum appearing in the definition of $C(K, \Omega)$ arbitrarily small already with *radial* functions u, and this is an elementary one-dimensional problem.

Putting the two facts above together and using the results of Section IX.1, we obtain the following

X.2.2 Theorem   *Let G be a connected Lie group of topological dimension n and let us assume that there exists a quasiregular mapping $\phi \colon \mathbb{R}^n \to G$. Then there exists $C > 0$ such that $V_G(t) \leq Ct^n$ $(t > 1)$.*

Quite a bit more can be said. First of all observe that there is no loss of generality in assuming in the above theorem that G is simply connected. Indeed we can lift ϕ to $\tilde{\phi} \colon \mathbb{R}^n \to \tilde{G}$, where $\tilde{G}$ is the universal covering group of G.

Observe next that when G is topologically homeomorphic to $\mathbb{R}^n$, the condition $V(t) \leq Ct^n$ can be verified only in very few cases. We must have $G \cong G_1 \rtimes G_2$, a semidirect product where G_1, G_2 are Euclidean groups and where G_2 acts on G_1 as rotations, i.e. elements of $SO(k)$ with $G_1 \cong \mathbb{R}^k$. These facts are not trivial and they are contained in the work of Y. Guivarc'h. From them it is easy to deduce the positive result that when $G \cong \mathbb{R}^n$ (topologically) and when $V(t) \leq Ct^n$ then G is in fact quasi-isometric to $\mathbb{R}^n$.

The above methods can probably be refined further to show that the above are the *only* cases where we can have a quasiregular $\phi \colon \mathbb{R}^n \to G$. We shall not pursue the matter further but leave this as an open problem.

X.3 Brownian motion on covering manifolds and random walks on groups

The problem of transience and recurrence of Brownian motion on covering manifolds was the original motivation for the theory developed in this book. In this section, we shall give a solution to this problem. We shall assume that the reader is familiar with the theory of covering manifolds.

Let M be some compact connected manifold and let $\widetilde{M}$ be the universal (simply connected) covering manifold of M. The fundamental group $\pi_1(M)$ acts then on $\widetilde{M}$. Let $H \triangleleft \pi_1(M)$ be some normal subgroup; the manifold $\widetilde{M}/H$ lies then between $\widetilde{M}$ and M in the sense that the canonical maps

$$\widetilde{M} \to \widetilde{M}/H \to M$$

are covering maps, the group $\Gamma = \pi_1(M)/H$ acts on $M_1 = M/H$, and $M_1/\Gamma = M$. The group Γ is called the deck transformation group of the normal covering $p_1 \colon M_1 \to M$. Recall that $\pi_1(M)$, hence also Γ, is finitely generated. Apart from that, there is no constraint whatsoever on Γ: indeed, any finitely generated discrete group can appear as a deck tranformation group. Let us now endow M with a Riemannian structure (M, g); this structure can then be lifted canonically to M_1 so that p_1 is a local isometry. We can also consider the Laplace-Beltrami operators Δ and Δ_1 of M and M_1 respectively and the corresponding heat diffusion semigroups $e^{-t\Delta}$ and $e^{-t\Delta_1}$ on M and M_1.

We wish to answer the following question: when is the diffusion generated by Δ_1 transient on M_1? One usually refers to that diffusion as the canonical Brownian motion on M_1. Equivalently, one can ask whether the semigroup $e^{-t\Delta_1}$ admits a finite Green function

$$\mathcal{G}(x, y) = \int_0^\infty h_t^1(x, y) \, dt, \text{ for } x \neq y,$$

where $h_t^1(x, y)$ is the kernel of $e^{-t\Delta_1}$. The above Green function, when it exists, is the fundamental solution of Δ_1 on M_1. So alternatively we can ask whether for some (or equivalently for all) $x \in M_1$ there exists

$$0 \leq u \in C^\infty(M_1 \setminus \{x\}) \cap L^1_{\text{loc}}(M_1)$$

such that we have $\Delta_1 u = -\delta_x$ in the distribution sense.

There is a well-known criterion due to Beurling and Deny that says that this question has a positive answer if and only if for some (or all) $K \subset\subset M_1$ (a compact subset of positive measure) we have $C_2(K; M_1) > 0$. Here C_2 denotes the 2-capacity of the condensor $K \subset\subset \Omega = M_1$. This is defined just like the conformal capacity (n-capacity) of Section IX.1 except that the power $|\nabla u|^n$ in the definition is replaced by $|\nabla u|^2$ (similarly, all the p-capacities $C_p(K, \Omega)$ can be defined with $p > 0$).

An immediate consequence of the above criterion is that the answer to our question is invariant under change of Riemannian metric on M, since two different metrics on M give rise to quasi-isometric metrics on M_1 and therefore the positivity of $C_2(K, M_1)$ is not altered.

Now let Γ be the deck transformation group of the covering $M_1 \to M$. We shall consider $\mu \in \mathbb{P}(\Gamma)$, a finitely supported symmetric probability measure on Γ that charges some fixed set of generators (i.e. such that $\mathrm{Gp}(\mathrm{Supp}\,\mu) = \Gamma$) and we shall ask ourselves whether the Markov chain given by the transition matrix $P(g, h) = \mu(g^{-1}h)$ $(g, h \in \Gamma)$ is transient. Equivalently, we shall ask whether the following series converges:

$$\sum_{k \geq 1} \mu^k(e) < +\infty. \tag{T}$$

$\mu^k = \mu * \mu * ... * \mu$ denotes here the kth convolution power of μ. We can consider the semigroup $\exp(-t(\delta - \mu))$ and it is an easy matter to verify that the series (T) converges if and only if the Green function is finite:

$$\mathcal{G}(x) = \int_0^\infty \exp(-t(\delta - \mu))(x)\, dt < +\infty, \quad x \in \Gamma.$$

The Beurling-Deny criterion applies therefore to this semigroup and it follows that $\mathcal{G}$ is finite if and only if

$$\inf_{d(x,y)=1} \sum |f(x) - f(y)|^2 > 0$$

where the infimum is taken over all $f \in C_0(\Gamma)$ with $f(e) \geq 1$. The distance $d(.,.)$ is of course the canonical word distance defined on the finitely generated group Γ that we considered in Chapter VI. The above Beurling-Deny criterion shows in particular that the convergence or divergence of the series (T) is independent of the particular choice of the measure μ. The groups for which the series (T) converges are called transient groups.

Another important consequence of the Beurling-Deny criterion is the following

X.3.1 Theorem　 *The Brownian motion on the normal covering manifold M_1, which we considered at the beginning of this section, is transient if and only if the deck transformation group Γ is transient.*

The proof relies on the discretization procedure presented in Section 3 below; the Beurling-Deny criterion does the rest.

We therefore have the answer to the original question of this section since we have seen in Chapter VI that the only finitely generated groups that are recurrent are the finite extensions of $\{0\}$, $\mathbb{Z}$ and $\mathbb{Z}^2$.

Now that we know exactly when a Green function exists on M_1, we could try to decide when there exist non-trivial positive or bounded harmonic functions. It is tempting to conjecture that this is the case exactly when there

exist non-trivial positive or bounded harmonic functions on the deck transformation group Γ. Here of course we use the definition that the function u
on Γ is harmonic with respect to a measure μ if $u * \mu = u$.

X.4 Dimension at infinity of a covering manifold

In this section we shall pursue the study of the heat semigroup $e^{-t\Delta_1}$ on
the covering manifold M_1. By the standard theory of elliptic operators,
that we take here for granted, it follows that the local dimension for both
$e^{-t\Delta}$ and $e^{-t\Delta_1}$ is n, the topological dimension of M. The semigroup $e^{-t\Delta}$
converges as $t \to \infty$ to the average operator over M for the volume measure
(again, by standard theorems). So, in this circle of ideas, what remains to
be determined is the dimension at infinity of $e^{-t\Delta_1}$.

Of course, there is nothing special about the Laplace-Beltrami operator.
In the above situation, we could have considered A, an elliptic (or even
subelliptic) second order differential operator on M that is formally self-
adjoint with respect to some smooth non-vanishing measure on M, and
we could have lifted A locally to M_1 to obtain A_1, and the corresponding
semigroup e^{-tA_1}. The problem of the dimension at infinity for e^{-tA_1} is the
same and the answer the same also. In general terms, the answer is as
follows: the dimension at infinity of any of the above semigroups e^{-tA_1} is
the same as the dimension (at infinity) of any of the convolution semigroups
on the discrete group Γ.

Let us be more specific; assume that for some symmetric finitely supported
probability measure $\mu \in \mathbb{P}(\Gamma)$, we have $\mu^k(\{e\}) = O(k^{-D/2})$. We are going
to see that $||e^{-t\Delta_1}||_{1\to\infty} = O(t^{-D/2})$ as $t \to +\infty$. (The L^p norms are
of course taken with respect to the Riemannian volume element on M_1.)
Conversely, if for some $D \geq 0$ we have

$$||e^{-t\Delta_1}||_{1\to\infty} = O(t^{-D/2}) \quad \text{as } t \to \infty,$$

then, for every symmetric probability measure with finite and generating
support, μ, we have $\mu^k(\{e\}) = O(k^{-D/2})$.

Let us sketch a proof of these facts, assuming for simplicity that $D > 2$.
Thanks to the theory developed in Chapters VI and VII, the condition

$$\mu^k(\{e\}) = O(k^{-D/2}), \quad \text{for one or every suitable } \mu \in \mathbb{P}(\Gamma), \qquad (1)$$

is equivalent to

$$||f||_{2D/(D-2)} \leq C \sum_{x \sim y} |f(x) - f(y)|^2, \quad \forall f \in C_0(\Gamma)$$

where $x \sim y$ means that xy^{-1} belongs to a fixed set of generators of Γ; here
suitable means symmetric, with a finite and generating support.

Fix now $x_0 \in M$, and let $X = p_1^{-1}(x_0)$. There is a natural identification
between X and Γ, and a natural connected graph structure on X: two points

of X are neighbours if they are associated with x and y in Γ such that $x \sim y$. The number of neighbours of a point X is bounded above (by the number of generators of Γ). Let d be the graph distance on X.

The condition (1) holds if and only if

$$\|f\|_{2D/(D-2)} \le CD(f), \quad \forall f \in C_0(X), \tag{2}$$

where

$$D(f)^2 = \sum_{d(x,y)=1} |f(x) - f(y)|^2.$$

Assume now that

$$\|e^{-t\Delta_1}\|_{1 \to \infty} = O(t^{-D/2}), \quad \text{as } t \to +\infty. \tag{3}$$

It follows from the fact that M is compact and that M_1 is a normal covering of M that

$$\exists c_2 \ge c_1 > 0 \text{ such that } \forall x, y \in X \quad x \ne y \Rightarrow d(x,y) \ge 2c_1$$
$$\forall x \in M, \quad d(x, X) \le c_2.$$

Now it is easy, using the fact that each point of X has a bounded number of neighbours, to construct a C_0^∞ partition of unity on M_1, $(\phi_x)_{x \in X}$, such that

$$\phi_x \equiv 1 \text{ in } B(x, c_1/4), \quad \text{supp}(\phi_x) \subset B(x, 4c_2), \quad \text{and } |\nabla \phi_x| \le C, \forall x \in X.$$

For $f \in C_0(X)$, set $\hat{f} = \sum f(x)\phi_x$. Then $\hat{f}$ belongs to $C_0^\infty(M_1)$, and for $y \in X$,

$$\nabla \hat{f} = \sum_{x \in X} [f(x) - f(y)] \nabla \phi_x.$$

Hence

$$|\nabla \hat{f}(t)| \le C \sup\{|f(x) - f(y)| \mid x \in X, d(x,y) \le 5c_2\}, \quad \forall t \in B(y, c_2),$$

since, for $t \in B(y, c_2)$, $\nabla \phi_x(t) \ne 0 \Rightarrow d(x,t) \le 4c_2 \Rightarrow d(x,y) \le 5c_2$. Now

$$\int |\nabla \hat{f}(t)|^2 \, d\sigma \le \sum_{x \in X} \int_{B(x,c_2)} |\nabla \hat{f}|^2 \, d\sigma \le C' \sum_{\substack{x,y \in X \\ x \sim_5 y}} |f(x) - f(y)|^2$$

$$\le 5C'K \sum_{x \sim y} |f(x) - f(y)|^2$$

where $x \sim_5 y$ means that x may be connected to y in at most five steps, and $K = \sup_{x \in X} \sharp\{y \in X \mid x \sim_5 y\}$. Therefore

$$\int |\nabla \hat{f}|^2 \, d\sigma \le CD(f)^2.$$

By Remark II.4.4 (a), (3) implies

$$\Delta_1^{-\frac{1}{2}} : L^2 \to L^2 + L^{2D/(D-2)}.$$

Therefore, if S is regularizing, $S\Delta^{-\frac{1}{2}}\colon L^2 \to L^{2D/(D-2)}$, hence

$$\|S\hat{f}\|^2_{2D/(D-2)} \le \|\Delta^{\frac{1}{2}}\hat{f}\|^2_2 = \int |\nabla \hat{f}|^2\, d\sigma \le CD(f)^2.$$

To obtain the inequality (2), therefore (1), it suffices to see that we can choose S such that

$$\|f\|_{2D/(D-2)} \le C\|S\hat{f}\|_{2D/(D-2)}.$$

It is easy to check that the operator

$$S\phi = \sum_{x\in X} \frac{1}{|B(x, c_1/4)|} \left(\int_{B(x, c_1/4)} \phi\, d\sigma \right) \phi_x$$

does the job.

Let us now show that $(1) \Rightarrow (3)$. For $x \in X$ and $\phi \in C_0^\infty(M_1)$, set

$$f(x) = \frac{1}{|B(x, 2c_2)|} \int_{B(x, 2c_2)} \phi\, d\sigma.$$

Poincaré's inequality tells us that

$$\int_{B(x, 2c_2)} |\phi(y) - f(x)|^2\, d\sigma(y) \le C \int_{B(x, 4c_2)} |\nabla \phi|^2\, d\sigma.$$

Now Jensen's inequality yields

$$\forall x, y \in X, x \sim y, \quad |f(x) - f(y)|^2 \le C \int_{B(x, 4c_2)} |\nabla \phi|^2\, d\sigma.$$

It follows that

$$D(f)^2 \le C \int |\nabla \phi|^2\, d\sigma.$$

Moreover, since (1) implies (2),

$$\|f\|_{2D/(D-2)} \le CD(f).$$

Now, if

$$\tilde{S}\phi = \sum_{x\in X} \frac{1}{|B(x, 2c_2)|} \left(\int_{B(x, 2c_2)} \phi\, d\sigma \right) \phi_x,$$

we have

$$\|\tilde{S}\phi\|_{2D/(D-2)} = \|\sum f(x)\phi_x\|_{2D/(D-2)} \le C\|f\|_{2D/(D-2)}.$$

Putting the above inequalities together gives

$$\|\tilde{S}\phi\|^2_{2D/(D-2)} \le C \int |\nabla \phi|^2\, d\sigma.$$

On the other hand, thanks again to Poincaré's inequality,

$$\|\tilde{S}\phi - \phi\|^2_2 \le C \int |\nabla \phi|^2\, d\sigma.$$

In other words $\Delta_1^{-\frac{1}{2}} = \tilde{S}\Delta_1^{-\frac{1}{2}} + (I - \tilde{S})\Delta_1^{-\frac{1}{2}} : L^2 \to L^{2D/(D-2)} + L^2$, which implies (3), by Remark II.4.4 (a).

Let us draw the consequences of the above considerations. If the volume growth function V of the deck transformation group Γ is such that $V(k) \geq k^D$, then $\|e^{-t\Delta_1}\|_{1\to\infty} = O(t^{-D/2})$ as $t \to \infty$. In particular, assume that n, the dimension of M, is at least $n \geq 2$, and that the growth function $V(k)$ of Γ grows "strictly faster than n". This is the case, because of Bass' and Gromov's theorems, as soon as we do *not* have $V(k) = O(k^n)$. From the above and Theorem II.4.3, it follows then that there exists some $p > 1$ such that

$$\|f\|_p \leq C\|\Delta_1^{\frac{1}{2}} f\|_n, \quad f \in C_0^\infty(M_1).$$

X.5 Quasiregular maps and compact manifolds

In this section, we shall prove the following

X.5.1 Theorem   *Let M be a compact Riemannian manifold of dimension n and assume that there exists $\phi : \mathbb{R}^n \to M$ a (non-constant) quasiregular map. Then the fundamental group of M is nilpotent and its growth function satisfies*

$$V(k) = O(k^n).$$

It is clear that we can assume that $n \geq 2$. It is also clear that the quasiregular map ϕ of the above theorem lifts to a map $\tilde{\phi} : \mathbb{R}^n \to \widetilde{M}$ on the universal covering space $\widetilde{M} \to M$ for which the deck transformation group is $\Gamma = \pi_1(M)$. The theorem will follow if we show that, as soon as we do not have $V_\Gamma(k) = O(k^n)$, the capacity of every condensor $K \subset\subset \Omega \subset \widetilde{M}$ is strictly positive

$$C(K;\Omega) > 0. \tag{C}$$

Indeed, we then just have to follow the proof given in Section X.2 verbatim. And just as there, it is therefore enough to prove that as soon as we do *not* have $V(k) = O(k^n)$, then there exists some $p \geq 1$ such that

$$\|f\|_p \leq C\|\nabla f\|_n; \quad f \in C_0^\infty(\widetilde{M}). \tag{S}$$

But, of course, this is almost the observation that we made at the end of Section X.3; almost, but not quite. What we need is the following "fact":

$$\|\nabla f\|_p \approx \|\Delta^{\frac{1}{2}} f\|_p, \quad f \in C^\infty(M_1).$$

Unfortunately, no proof of this exists; indeed, in this generality (i.e. for any covering of a compact manifold), it is likely to be difficult to show. This "fact" is sometimes referred to as the "M. Riesz theorem" for the corresponding Riemannian manifold, because it *is* the M. Riesz theorem when $M_1 \cong \mathbb{R}^n$.

However, there is another procedure for obtaining the estimate (S) or (C) that avoids the use of the "M. Riesz theorem" for Riemannian manifolds. This consists in proving first the analogue of (S) or (C) directly on the fundamental group Γ.

For any $f \in c_0(\Gamma)$ let us denote by

$$\nabla f(x) = \sum_j |f(x) - f(x\gamma_j)|$$

where $\gamma_1, ..., \gamma_s$ is some fixed set of generators. Under the assumption that $V(k) = O(k^n)$ does *not* hold, we wish to show that

$$\|f\|_p \leq C\|\nabla f\|_n; \quad f \in C_0(\Gamma) \tag{S}$$

for some $p \geq 1$, $C > 0$. From this we can deduce (S) on M_1 by discretization or better we can deduce directly (because after all this is all that is needed) that

$$C(K, \Omega) > 0$$

for any condensor $K \subset\subset \Omega \subset \widetilde{M}$.

We are still left with the proof of the estimate (S). When Γ is not virtually nilpotent this inequality follows from Theorem VI.3.1. When Γ is virtually nilpotent, the answer is given by Theorem VI.5.3.

References and comments

The background material on differential geometry is classical. The results on quasiregular mappings that we use can be found in [1], [102], [124], [156]. The problems treated in Sections X.1 and X.2 stem from [57], [102]. These results have not been published elsewhere and are due to Varopoulos. For results concerning harmonic functions, see [87], [60], [83], [82], [147], [2], [3], [4], [63], [114], and the references cited in these papers. Background and developments on covering manifolds and Green function can be found in [1] and [9].

For Section 4, see [128], [131], [133], [146], [72], [28], [29], [31], [114]. In Section 5, we encountered the problem of whether on a Riemannian manifold we have

$$\|\nabla f\|_p \approx \|\Delta^{\frac{1}{2}} f\|_p, \quad f \in C_0^\infty.$$

The same problem can of course be considered on a (unimodular) Lie group G where ∇ and Δ are invariant under the group structure but need not necessarily be Riemannian (but instead have the meaning given them throughout in this book). The general problem still remains open; see however References and Comments of Chapter VIII.

BIBLIOGRAPHY

[1]   L. Ahlfors. *Conformal invariants*, McGraw-Hill, New-York, 1979.

[2]   G. Alexopoulos. Fonctions harmoniques bornées sur les groupes résolubles, *C. R. Acad. Sci. Paris*, **305**, 1987, 777-779.

[3]   G. Alexopoulos. On the mean distance of random walks on groups, *Bull. Sc. Math.*, 2ème série, **111**, 1987, 189-199.

[4]   G. Alexopoulos. Quelques propriétés des fonctions harmoniques sur les groupes de Lie à croissance polynômiale, *C. R. Acad. Sci. Paris*, **308**, 1989, 337-338.

[5]   G. Alexopoulos. Inégalités de Harnack paraboliques et transformées de Riesz sur les groupes de Lie résolubles à croissance polynômiale du volume, *C. R. Acad. Sci. Paris*, **309**, 1989, 661-662.

[6]   G. Alexopoulos, N. Lohoué. Sobolev inequalities and harmonic functions of polynomial growth, preprint.

[7]   G. Alexopoulos. An application of homogenisation theory to harmonic analysis: Harnack inequalities and Riesz transforms on Lie groups of polynomial growth, to appear in *Can. J. Math.*

[8]   G. Alexopoulos. A lower estimate for central probabilities on polycyclic groups, preprint.

[9]   A. Ancona. Théorie du potentiel sur des graphes et des variétés, *Cours de l'Ecole d'été de probabilités de Saint-Flour*, Springer L. N. 1427, 1988.

[10]   D. Aronson. Bounds for the fundamental solution of a parabolic equation, *Bull. of the A.M.S.*, **73**, 1967, 890-896.

[11]   P. Baldi. Caractérisation des groupes de Lie connexes récurrents, *Ann. Inst. H. Poincaré, prob. et stat.*, **17**, 1981, 281-308.

[12]   P. Baldi, N. Lohoué, J. Peyrière. Sur la classification des groupes récurrents, *C. R. Acad. Sci. Paris*, **285**, série I, 1977, 1103-1104.

[13]   H. Bass. The degree of polynomial growth of finitely generated nilpotent groups, *Proc. London Math. Soc.*, **25**, 1972, 603-614.

[14]   P. Bénilan. Opérateurs accrétifs et semi-groupes dans les espaces L^p $(1 \leq p \leq +\infty)$, in *Functional Analysis and Numerical Analysis*, Japan-France seminar, H. Fujita (ed.), Japan Society for the Advancement of Science, 1978.

[15]   J.-M. Bony. Principe du maximum, inégalité de Harnack et unicité du problème de Cauchy pour les opérateurs elliptiques dégénérés, *Ann. Inst. Fourier*, **19**, 1969, 277-304.

[16]   P. Bougerol. Théorème central limite local sur certains groupes de Lie, *Ann. Sc. E. N. S.*, série 4, **14**, 1981, 403-432.

[17]   P. Buser. A note on the isoperimetric constant, *Ann. Scien. Ec. Norm. Su* , **15**, 1982, 213-230.

[18]   P. Buser, H. Karcher. Gromov's almost flat manifolds, *Astérisque*, **81**, 1981.

[19] C. Carathéodory. Untersuchungen über die Grundlagen der Thermo-dynamik, *Math. Ann.*, **67**, 1909, 355-386.

[20] E. Carlen, S. Kusuoka, D. Stroock. Upper bounds for symmetric Markov transition functions, *Ann. Inst. H. Poincaré, prob. et stat.*, **23**, 1987, 245-287.

[21] K. Carne. A transmutation formula for Markov chains, *Bull. Sc. Math.*, 2ème série, **109**, 1985, 399-403.

[22] W. Chow. Über Systemen von linearen Partiellen Differentialgleichun-gen erster Ordnung, *Math. Ann.*, **117**, 1939, 98-105.

[23] M. Christ. L^p bounds for spectral multipliers on nilpotent groups *Trans. Am. Math. Soc.*, **328**, 1991, 73-81.

[24] T. Coulhon. Dimension à l'infini d'un semi-groupe analytique, *Bull. Sc. Math.*, 2ème série, **114**, 1990, 485-500.

[25] T. Coulhon. Dimensions of continuous and discrete semigroups, in *Semigroup theory and evolution equations*, Clément, Mitidieri, De Pagter, eds., Marcel Dekker L. N. in Pure and Appl. Math., **135**, 1991, 93-99.

[26] T. Coulhon. Itération de Moser et estimation gaussienne du noyau de la chaleur, to appear in *Jour. Oper. Th.*

[27] T. Coulhon. Noyau de la chaleur et discrétisation d'une variété rie-mannienne, to appear in *Israël. J. Math.*

[28] T. Coulhon. Inégalités de Gagliardo-Nirenberg pour les semi-groupes d'opérateurs et applications, preprint.

[29] T. Coulhon. Sobolev inequalities on graphs and on manifolds, to appear in the proceedings of the conference *Harmonic Analysis and Discrete Potential Theory*, Frascati, 1991.

[30] T. Coulhon, D. Lamberton. Quelques remarques sur la régularité L^p du semi-groupe de Stokes, *Comm. in P.D.E.*, **17**, 1992, 287-304

[31] T. Coulhon, M. Ledoux, Un résultat négatif concernant l'isopérimét-rie, preprint.

[32] T. Coulhon, L. Saloff-Coste. Théorie de Hardy-Littlewood-Sobolev pour les semi-groupes d'opérateurs et application aux groupes de Lie uni-modulaires, in *Séminaire d'analyse de l'Université de Clermont-Ferrand II, 1987-88*, exposé no. 21, 1989.

[33] T. Coulhon, L. Saloff-Coste. Théorèmes de Sobolev pour les semi-groupes d'opérateurs et application aux groupes de Lie unimodulaires, *C. R. Acad. Sci. Paris*, **309**, série I, 1989, 289-294.

[34] T. Coulhon, L. Saloff-Coste. Semi-groupes d'opérateurs et espaces fonctionnels sur les groupes de Lie, *Jour. Approx. Th.*, **65** 1991, 176-199.

[35] T. Coulhon, L. Saloff-Coste. Puissances d'un opérateur régularisant, *Ann. Inst. H. Poincaré, prob. et stat.*, **26**, 1990, 419-436.

[36] T. Coulhon, L. Saloff-Coste. Marches aléatoires non symétriques sur les groupes unimodulaires, *C. R. Acad. Sci. Paris*, **310**, série I, 1990, 627-630.

[37]   T. Coulhon, L. Saloff-Coste. Isopérimétrie pour les groupes et les variétés, preprint.

[38]   M. Cowling, S. Meda. Harmonic analysis and ultra contractivity, to appear in *Trans. A.M.S.*.

[39]   E. B. Davies. *One parameter semi-groups*, Academic press, 1980.

[40]   E. B. Davies. Explicit constants for gaussian upper bounds on heat kernels, *Amer. J. Math.*, **109**, 1987, 319-334.

[41]   E. B. Davies. Gaussian upper bounds for the heat kernels of some second order operators on Riemannian manifolds, *J. Funct. Anal.*, **80**, 1988, 16-32.

[42]   E. B. Davies. *Heat kernels and spectral theory*, Cambridge University Press, 1989.

[43]   E. B. Davies, M. Pang. Sharp heat kernel bounds for some Laplace operators, *Quart. J. Math. Oxford*, série 2, **40**, 1989, 281-290.

[44]   E. B. Davies, B. Simon. Ultracontractivity and the heat kernel for Schrödinger operators and Dirichlet Laplacians, *J. Funct. Anal.*, **59**, 1984, 335-395.

[45]   J. Deny. Méthodes hilbertiennes en théorie du potentiel, in *Potential theory*, C.I.M.E., Stresa, 1969.

[46]   J. Doob. *Potential theory and its probabilistic counterpart*, Springer Verlag, 1984.

[47]   E. Fabes, D. Stroock. A new proof of Moser's parabolic Harnack inequality via the old ideas of Nash, *Arch. Rat. Mech. Anal.*, **96**, 1986, 327-338.

[48]   C. Fefferman, S. Sanchez-Calle. Fundamental solutions for second order subelliptic operators, *Ann. Math.* **124**, 1986, 247-272.

[49]   G. Folland, E. Stein. *Hardy spaces on homogeneous groups*, Princeton University Press, 1982.

[50]   M. Fukushima. *Dirichlet forms and Markov processes*, North Holland, 1980.

[51]   M. Fukushima. On an L^p estimate of resolvents of Markov processes, *Publ. R.I.M.S., Kyoto Univ.*, **13**, 1977, 277-284.

[52]   E. Gagliardo. Proprieta di alcune classi di funzioni in piu variabili, *Ricerche Mat.*, **7**, 1953, 102-137.

[53]   N. Goodman. *Nilpotent Lie groups*, Springer Lecture Notes 562, 1976.

[54]   R. Grigorchuk. Degrees of growth of finitely generated groups and the theory of invariant means, *Izv. Akad. Nauk. SSSR Ser. Math* **48**, 1984, 939-985 (Russian); *Math. USSR Izvestiya*, **25**, 1985, 259-300 (English translation).

[55]   R. Grigorchuk. *Proceedings of the Kyoto I.C.M.*, 1990.

[56]   M. Gromov. Groups of polynomial growth and expanding maps, *Publ. Math. I.H.E.S.*, 53, 1981, 53-78.

[57]   M. Gromov, *Structures métriques pour les variétés Riemanniennes*, Cedic-Nathan, 1981.

[58]   L. Gross. Logarithmic Sobolev inequalities, *Amer. J. Math.*, **97**, 1976, 1061-1083.

[59]   Y. Guivarc'h. Croissance polynômiale et périodes des fonctions harmoniques, *Bull. Soc. Math. France*, **101**, 1973, 333-379.

[60]   Y. Guivarc'h. Mouvement brownien sur les revêtements d'une variété compacte, *C. R. Acad. Sc. Paris*, série I, **292**, 1981, 851-853; and **101**, 1973, 333-379.

[61]   Y. Guivarc'h, M. Keane, B. Roynette, *Marches aléatoires sur les groupes de Lie*, Springer Lecture Notes 624, 1977.

[62]   W. Hebisch. On heat kernels on Lie groups, to appear in *Math. Z.*

[63]   W. Hebisch, L. Saloff-Coste. Gaussian estimates for Markov chains and random walks on groups, preprint.

[64]   S. Helgason. *Differential Geometry, Lie Groups, and Symmetric Spaces*, Academic Press, 1978.

[65]   L. Hörmander. Hypoelliptic second-order differential equations, *Acta. Math.*, **119**, 1967, 147-171.

[66]   L. Hörmander, A. Melin. Free systems of vector fields, *Arkiv för Math.*, **16**, 1978, 83-88.

[67]   A. Hulanicki, J. Jenkins. Nilpotent Lie groups and summability of eigenfunction expansions of Schrödinger operators, *Studia Math.*, **80**, 1984, 235-244.

[68]   J. Jenkins. Growth of connected locally compact groups, *J. Funct. Anal.*, **12**, 1973, 113-127.

[69]   D. Jerison, J. Lee, Extremals for the Sobolev inequality on the Heisenberg group and the CR Yamabe problem, *J. Amer. Math. Soc.*, **1**, 1988, 1-13.

[70]   D. Jerison, A. Sanchez-Calle. Estimates for the heat kernel for a sum of squares of vector fields, *Indiana Univ. Math. J.*, **35**, 1986, 835-854.

[71]   D. Jerison, A. Sanchez-Calle, *Subelliptic second order differential operators*, Springer Lecture Notes 1277, 1986.

[72]   M. Kanai. Analytic inequalities, and rough isometries between noncompact Riemannian manifolds, Springer Lecture Notes 1201, 1986, 122-137.

[73]   H. Kesten. The Martin boundary of recurrent random walks on countable groups, in *Proc. 5th Berkeley Sym on Math. Statistics and Probability*, vol. II, 1967, 51-74.

[74]   H. Kesten. Symmetric random walks on groups, *Trans. A.M.S.*, **92**, 1959, 336-354.

[75]   H. Komatsu. The Sobolev-Besov imbedding theorem from the point of view of semigroups of operators, *Séminaire Goulaouic-Schwartz*, exposé 1, 1972, 1973.

[76]   N. Krylov, M. Safonov. A certain property of solutions of parabolic equations with measurable coefficients. *Math. USSR-Izv.* 16, 1981, 151-164.

[77]   N. Lohoué. Estimées de type Hardy pour l'opérateur $\Delta + \lambda$ d'un espace symétrique de type non compact, *C. R. Acad. Sci. Paris*, **308**, série

I, 1989, 11-14.

[78]   N. Lohoué. Variantes des inégalités de Sobolev et inégalités de Trudinger pour les groupes de Lie et les variétés riemanniennes, *Forum mathematicum*, **3**, 1991, 371-388.

[79]   N. Lohoué. Estimées de type Hardy-Sobolev sur certaines variétés riemanniennes et les groupes de Lie, preprint.

[80]   N. Lohoué. Inégalités de Sobolev pour les sous-laplaciens de certains groupes unimodulaires, preprint.

[81]   N.Lohoué, N. Varopoulos. Remarques sur les transformées de Riesz sur les groupes de Lie nilpotents, *C. R. Acad. Sci. Paris*, **301**, série I, 1985, 559-560.

[82]   T. Lyons. Instability of the Liouville property for quasi-isometric Riemannian manifolds and reversible Markov chains. *J. Diff. Geom.* **26**, 1987, 33-66.

[83]   T. Lyons, D. Sullivan. Function theory, random paths and covering spaces, *J. Diff. Geom.* **19**, 1984, 299-323.

[84]   P. Maheux. Analyse et géométrie sur les espaces homogènes, preprint.

[85]   P. Maheux. Analyse et géometrie sur les espaces homogènes, Thèse de doctorat, Université Paris VI.

[86]   A. Mal'cev. On a class of homogeneous spaces, *Amer. Math. Soc. Translations*, **39**, 1951, 276-307.

[87]   G. Margulis. Positive harmonic functions on nilpotent groups. *Sov. Math. Doklady*, **7**, 1966, 241-244.

[88]   G. Mess. Varopoulos's theorem on transient groups, *M.S.R.I. Publ.*, 1989.

[89]   J. Milnor. A note on curvature and fundamental group, *J. Diff. Geom.*, **2**, 1968, 1-7.

[90]   J. Moser. On Harnack's theorem for elliptic differential equations. *Comm. Pure Appl. Math.*, **14**, 1961, 557-591.

[91]   J. Moser. A Harnack inequality for parabolic differential equations, *Comm. Pure Appl. Math.*, **17**, 1964, 101-134.

[92]   J. Moser. Correction to "A Harnack inequality for parabolic differential equations", *Comm. Pure Appl. Math.*, **20**, 1967, 232-236.

[93]   J. Moser. On pointwise estimates for parabolic differential equations, *Comm. Pure Appl. Math.*, **24**, 1971, 727-740.

[94]   A. Nagel, E. Stein and M. Wainger. Balls and metrics defined by vector fields, *Acta. Math.*, **155**, 1985, 103-147.

[95]   J. Nash. Continuity of solutions of parabolic and elliptic equations, *Amer. J. Math.*, **80**, 1958, 931-954.

[96]   L. Nirenberg. On elliptic partial differential equations, *Ann. Sc. Norm. Su Pisa*, **13**, 1959, 116-162.

[97]   P. Pansu. Une inégalité isopérimétrique sur le groupe de Heisenberg, *C. R. Acad. Sci. Paris*, **295**, série I, 1982, 127-130.

[98]   A. Pazy. *Semigroups of linear operators and applications to partial differential equations*, Springer Verlag, 1983.

[99]   M. Ragunathan. *Discrete subgroups of Lie groups.* Springer Verlag, 1972.

[100]   A. Raugi. Théorème de la limite centrale sur les groupes de Lie nilpotents, *Z. Wahrs. Verw. Geb.*, **43**, 1978, 149-172.

[101]   A. Raugi. Théorème de la limite centrale pour un produit semi-direct d'un groupe de Lie résoluble simplement connexe de type rigide par un groupe compact, in *Probability measures on groups*, Springer L.N. Math. 706, 1978, 257-324.

[102]   S. Rickman, Forthcoming book on Quasiregular Mappings, Springer Verlag.

[103]   D. Robinson. *Elliptic Operators and Lie groups*, Oxford University Press, 1991.

[104]   L. Rothschild, E. Stein. Hypoelliptic differential operators and nilpotent groups, *Acta Math.*, **137**, 1976, 247-320.

[105]   L. Saloff-Coste. Analyse sur les groupes de Lie nilpotents, *C. R. Acad. Sci. Paris*, **302**, série I, 1986, 499-502.

[106]   L. Saloff-Coste. Fonctions maximales sur certains groupes de Lie, *C. R. Acad. Sci. Paris*, **305**, série I, 1987, 457-459.

[107]   L. Saloff-Coste. Théorèmes de Sobolev et inégalités de Trudinger sur certains groupes de Lie, *C. R. Acad. Sci. Paris*, **306**, série I, 1988, 305-308.

[108]   L. Saloff-Coste. Inégalités de Sobolev produit sur les groupes de Lie nilpotents, *J. Funct. Anal.*, **79**, 1988, 44-56.

[109]   L. Saloff-Coste. Sur la décroissance des puissances de convolution sur les groupes, *Bull. Sc. Math.*, 2ème série, **113**, 1989, 3-21.

[110]   L. Saloff-Coste. Sobolev inequalities and polynomial decay of convolution powers and random walks, to appear in the *Proceedings of the 1989 Summer Conference on Stochastic Analysis*, Lisbon.

[111]   L. Saloff-Coste. Analyse réelle sur les groupes à croissance polynômiale, *C. R. Acad. Sci. Paris*, **309**, série I, 1989, 149-151.

[112]   L. Saloff-Coste. Analyse sur les groupes de Lie à croissance polynômiale, *Ark. för Math.*, **28**, 1990, 315-331.

[113]   L. Saloff-Coste. Uniformly elliptic operators on Riemannian manifolds, to appear in *J. Diff. Geom.*

[114]   L. Saloff-Coste. A note on Poincaré, Sobolev and Harnack inequalities, to appear in *I. M. R. N. Duke Math. J.*

[115]   L. Saloff-Coste, D. Stroock. Opérateurs uniformément sous-elliptiques sur les groupes de Lie *J. Funct. Anal.*, **98**, 1991, 97-121.

[116]   A. Sanchez-Calle. Fundamental solutions and geometry of the sum of squares of vector fields. *Invent. Math.* **78**, 1984.

[117]   S. Sobolev. On a theorem of functional analysis. *A. M. S. translations*, série 2, **34**, 1963, 39-68.

[118] E. Stein. *Singular integrals and differentiability properties of functions*, Princeton University Press, 1970.

[119] E. Stein. *Topics in harmonic analysis related to the Littlewood-Paley theory*, Princeton University Press, 1970.

[120] D. Stroock, S. Varadhan, *Multidimensional diffusion processes*, Springer Verlag, 1979.

[121] J. Tits. Free subgroups of linear groups, *J. Algebra*, **20**, 1972, 250-270.

[122] J. Tits. Groupes à croissance polynômiale (d'aprés Gromov et al.). Springer Lecture Notes 901, 1981, 176-188.

[123] F. Trèves. *Pseudodifferential and Fourier integral operators*, vol.1, Plenum, New York, 1980.

[124] M. Vaisala, *Lectures on n-dimensional quasi-conformal mappings*, Springer Lecture Notes 229, 1971.

[125] L. Van de Dries, A. Wilkie. Gromov's theorem on groups of polynomial growth and elementary logic. *J. of Alg.* **89**, 1984, 349-374.

[126] V. S. Varadarajan. Lie groups, Lie algebras and their representations, Springer Verlag, 1984.

[127] N. Varopoulos. Random walks on soluble groups. *Bull. Sc. Math.* 2ème série, **107**, 1983, 337-344.

[128] N. Varopoulos. Brownian motion and transient groups. *Ann. Inst. Fourier.* 3, 1983, 241-261.

[129] N. Varopoulos. Chaînes de Markov et inégalités isopérimétriques, *C. R. Acad. Sci. Paris*, **298**, série I, 1984, 233-236.

[130] N. Varopoulos. Chaînes de Markov et inégalités isopérimétriques, *C. R. Acad. Sci. Paris*, **298**, série I, 1984, 465-468.

[131] N. Varopoulos. Brownian motion and random walks on manifolds, *Ann. Inst. Fourier*, **34**, 1984, 243-269.

[132] N. Varopoulos. Une généralisation du théorème de Hardy–Littlewood–Sobolev pour les espaces de Dirichlet, *C. R. Acad. Sci. Paris*, **299**, série I, 1984, 651-654.

[133] N. Varopoulos. Brownian motion can see a knot. *Math. Proc. Camb. Phil. Soc.* **97**, 1985, 299-309.

[134] N. Varopoulos. Long range estimates for Markov chains, *Bull. Sc. Math.*, 2ème série, **109**, 1985, 225-252.

[135] N. Varopoulos. Isoperimetric inequalities and Markov chains, *J. Funct. Anal.*, **63** 1985, 215-239.

[136] N. Varopoulos. Hardy-Littlewood theory for semi-groups, *J. Funct. Anal.*, **63** 1985, 240-260.

[137] N. Varopoulos. A potential theoretic property of groups. *Bull. Sc. Math.*, 2ème série, **109**, 1985, 113-119.

[138] N. Varopoulos. Théorie du potentiel sur les groupes nilpotents, *C. R. Acad. Sci. Paris*, **301**, série I, 1985, 143-144.

[139] N. Varopoulos. Itération de Moser, Perturbation de semi-groupes sous-markoviens. *C. R. Acad. Sci. Paris*, **301**, série I, 1985, 617-620.

[140] N. Varopoulos. Semi-groupes d'opérateurs sur les espaces L^p. *C. R. Acad. Sci. Paris*, **301**, série I, 1985, 865-868.

[141] N. Varopoulos. Théorie du potentiel sur des groupes et des variétés, *C. R. Acad. Sci. Paris*, **302**, série I, 1986, 203-205.

[142] N. Varopoulos. L'analyse sur les groupes de Lie unimodulaires. *C. R. Acad. Sci. Paris*, **302**, série 1, 1986, 503-506.

[143] N. Varopoulos. Analyse sur les groupes unimodulaires. *C. R. Acad. Sci. Paris*, **303**, série I, 1986, 93-95.

[144] N. Varopoulos. Analysis on nilpotent groups, *J. Funct. Anal.*, **66**, 1986, 406-431.

[145] N. Varopoulos. Convolution powers on locally compact groups, *Bull. Sc. Math.*, 2ème série, **111**, 1987, 333-342.

[146] N. Varopoulos. Random walks and Brownian motion on manifolds, *Symposia Mathematica*, **XXIX**, 1987, 97-109.

[147] N. Varopoulos. Fonctions harmoniques sur les groupes de Lie, *C. R. Acad. Sci. Paris*, **304**, série I, 1987, 519-521.

[148] N. Varopoulos. Analyse sur les groupes de Lie non unimodulaires, *C. R. Acad. Sci. Paris*, **305**, série I, 1987, 717-719.

[149] N. Varopoulos. Analysis on Lie groups, *J. Funct. Anal.*, **76**, 1988, 346-410.

[150] N. Varopoulos. Analyse sur les espaces symétriques et les groupes de Lie, *C. R. Acad. Sci. Paris*, **306**, série I, 1988, 115-116.

[151] N. Varopoulos. Sobolev inequalities on Lie groups and symmetric spaces, *J. Funct. Anal.*, **86**, 1989, 19-40.

[152] N. Varopoulos. Small time gaussian estimates of heat diffusion kernels. Part I. The semi-group technique, *Bull. Sc. Math.*, 2ème série, **113**, 1989, 253-277.

[153] N. Varopoulos. Small time gaussian estimates of heat diffusion kernels. II. The theory of large deviations, *J. Funct. Anal.*, **93**, 1990, 1-33.

[154] N. Varopoulos. Analysis and geometry on groups, *Proceedings of the Kyoto I.C.M.*, 1990.

[155] N. Varopoulos. Groups of superpolynomial growth, in *Proceedings of the I.C.M. satellite conference on Harmonic Analysis*, Springer Verlag, 1991.

[156] M. Vuosinen, *Conformal geometry and quasi-regular mappings*, Springer Lecture Notes 1319, 1988.

[157] J. Wolf. Growth of finitely generated solvable groups and curvature of Riemannian manifolds, *J. Diff. Geom.*, **2**, 1968, 421-446.

[158] A. Yoshikawa. Fractional powers of operators, interpolation theory and embedding theorems. *J. Fac. Sci. Univ. Tokyo*, IA, **18**, 1971.

[159] K. Yosida. *Functional Analysis*, Springer-Verlag, 1980.

INDEX